Fauna of New Zealand

Fauna of New Zealand

Number 9

Protura (Insecta)

S. L. Tuxen†

Department of Entomology, Zoological Museum

University of Copenhagen, Denmark*

†Deceased 15 June 1983. Contribution completed and submitted by Henrik Enghoff (see Foreword)

*Postal address: Zoologisk Museum, Universitetsparken 15, DK 2100 København

Cataloguing-in-publication citation

TUXEN, S. L.

Protura (Insecta) / S. L. Tuxen. – Wellington : DSIR, 1985.
(Fauna of New Zealand, ISSN 0111–5383; 9)
ISBN 0–477–06765–4

I. Title   II. Series

UDC 595.71

Date of publication: see back cover of Number 10

Suggested form of citation

Tuxen, S. L. 1985: Protura (Insecta). *Fauna of New Zealand [no.] 9.*

—⚘—

This publication was produced by offset lithography. The author's text was supplied as word processor output stored on diskette, and after editing and style coding was phototypeset at the N.Z. Government Printing Office. Times New Roman type is used for most of the text; major headings and figure labels are set in Univers; Garamond and Geneva are used in the titles.

The Editorial Advisory Group and the Series Editor acknowledge the following co-operation.

DSIR, Mt Albert Research Centre:

Mr D.W. Helmore – front cover motif, figure labelling

Mrs H.A. Whelan – computer operation and file management

Science Information Publishing Centre, DSIR:

Dr N. Hawcroft – style coding and phototypesetting

—⚘—

Front cover: The proturan depicted is *Amphientulus zelandicus* Tuxen n.sp.

Published by Science Information Publishing Centre, DSIR
P.O. Box 9741, Wellington, New Zealand

FOREWORD

Dr Søren Ludvig Tuxen died suddenly on 15 June 1983. Born in Copenhagen on 8 August 1908, he graduated from the University of Copenhagen in 1933, and received his doctoral degree in 1944 for a thesis about the hot springs of Iceland and their fauna. His main field of research has been the apterygote insects, in particular the Protura, as evidenced by the list of references in this contribution. He was employed at the Zoological Museum, Copenhagen, between 1933 and his retirement in 1978 (from 1958 to 1978 as Keeper of Entomology), and had continued his work at the museum up until his death.

The manuscript on Protura for *Fauna of New Zealand* was almost ready for submission at the time of Dr Tuxen's death; the text was written, and pencil draw ings were made. It fell to me to see to the completion of this important contribution to proturology. Dr Tuxen's drawings were inked by Mr Robert Nielsen, and the very extensive labelling was done by Mrs Hanne Jacobsen. Dr G. Imadaté, of Tokyo, took upon himself the substantial task of reviewing the completed manuscript and drawings, and furthermore provided diagnoses of supraspecific taxa other than the genera described as new. I want to express my gratitude to these persons, without whose help it would not have been possible to publish this work. The list of references and the figure captions were prepared by me. The publication of this work was supported by a grant, J.nr. 91–3068, from the Danish Natural Science Research Council.

It was Dr Tuxen's explicit wish that the illustrations be placed with the species descriptions to which they relate, contrary to the conventions adopted in the *Fauna of New Zealand.* I am grateful to the Series Editor for his willingness to accept this deviation.

HENRIK ENGHOFF
Chairman, Department of Entomology
Zoological Museum, University of Copenhagen

ABSTRACT

Sixteen species of Protura in ten genera are recognised from New Zealand, of which seven species and two genera are new to science. Fifteen species are fully described and illustrated; the sixteenth, known from a single immature specimen, is of indeterminate identity. Keys are given to the families of Protura and to the species known from New Zealand. The morphology and diagnostic characters of proturans are reviewed in some detail, and illustrated with scanning electron micrographs, as a further aid to identification. The known distribution of the species recorded here, and their systematic affinities, suggest a Gondwana origin for all but a few that are clearly recent introductions from Europe. Brief mention is made of the history of study of Protura in New Zealand, their life cycle and biology, and techniques used in their collection, preparation, and study.

CHECKLIST OF TAXA

CONTENTS

ACKNOWLEDGMENTS

I am very grateful to Dr G. W. Ramsay (NZAC), Mr R. G. Ordish (NMNZ), and Dr R. M. Emberson (LCNZ) for entrusting the specimens from their collections to my care for this study. Mr R. A. Savill (CMNZ) kindly lent me the lectotype of *Eosentomon dawsoni*. I owe thanks also to Mr Bent W. Rasmussen (ZMKD) for preparing the scanning electron micrographs presented as Figures 2–13.

INTRODUCTION

Before this study was undertaken, only very little was known about the proturan fauna of New Zealand. R. J. Tillyard (1925) was the first to mention the group in a New Zealand journal. He made a curious error, however, confounding Genova (Genoa) in Italy with Guinea in West Africa, even to mentioning the species *Acerentomon doderoi* Silvestri under the name *guineense*, which later caused some confusion. He supposed the Protura to be present in New Zealand, perhaps existing as especially large forms, but did not mention any findings.

The next author to refer to New Zealand Protura was G. H. Satchell (1952), who in a short note mentioned that B. J. Marples had collected in the Otago Museum garden some ten specimens belonging to the family Acerentomidae. I have not been able to locate these specimens. In the same year B. Condé described *Eosentomon dawsoni* from Little Barrier Island, on the basis of one female and one juvenile specimen. They were redescribed by Tuxen (1964, p. 123), and were studied again for this contribution.

Finally, H. Pauline McColl (1975) mentioned thirteen specimens from Kaitoke, near Wellington, but I have not been able to locate these either.

The present survey is based on a large holding of material in the New Zealand Arthropod Collection, Auckland; on twenty-nine specimens from the National Museum of New Zealand, Wellington; and on forty-two slides from the Department of Entomology, Lincoln College, Canterbury.

SYSTEMATICS AND PHYLOGENY

Protura are very small creatures, from less than 0.5 mm to 2 mm in length, white and softly sclerotised or harder and with a yellow tint, especially towards the hind end. Exceptionally they may be strongly sclerotised and red-brown in colour (*Sinentomon* from China, Korea, and Japan). The body is divided into three tagmata: a head with entognathous mouthparts, but without antennae or eyes; a three-segmented thorax with walking legs; and a twelve-segmented abdomen with rudimentary limb-appendages on the first three segments. This arangement of tagmata — head, three-segmented thorax, and abdomen of eleven or twelve segments — is characteristic of insects.

The lack of antennae has "caused" the forelegs to take over their sensory function, and they are furnished with sensory hairs (sensillae) the shape, number, and arrangement of which are of great systematic importance. On the supposed position of the eyes some organs called pseudoculi are situated; these may be homologous with the postantennal organs of Collembola and some Diplura. Spiracles are present on the mesothorax and metathorax in Eosentomoidea and *Sinentomon*, but the tracheae originating from each spiracle do not unite. The abdomen's full complement of eleven segments and a telson is not attained until maturation, a phenomenon (anamorphosis) that is exceptional among insects though usual in the myriapods. The external genitalia open behind the eleventh segment in both sexes, and are characteristically shaped. The spermatozoa are different from any other insect spermatozoon and very different in the two suborders of Protura (Baccetti *et al.* 1973).

The basic systematics of the group proposed by Tuxen (1964) is still that on which new additions are based. It is apparent in the key to suborders and families given on page 15.

Whether or not Protura are insects has long been, and is still, a subject of contention. In naming and erecting the order in 1907 Silvestri was of the opinion that they are insects, but in his admirable monograph of 1909 Berlese gave them the name Myrientomata, thus stressing their affinity to myriapods. The most recent view (and, incidentally, my own) is to regard them as a sister-group to Collembola within the subclass Entognatha (Hennig 1981); this in turn is a sister-group to Ectognatha, which comprises the Archaeognatha, Zygentoma (formerly Thysanura), and Pterygota (Tuxen 1980). As a sister-group to Collembola, which are known as fossils (*Rhyniella*) from Devonian strata, the Protura must be of the same antiquity at least. This fact I have used in my interpretation of their zoogeography (Tuxen 1978a).

ZOOGEOGRAPHY

The zoogeographical considerations just mentioned (Tuxen 1978a) were based on the theory of continental drift in its more recent guise of plate tectonic theory. I took the distribution of *Delamarentulus tristani* (Silvestri) in northern South America and western Africa, but nowhere else, as a starting point and based my conclusions as to the distribution of Protura on three suppositions.

(1) That the Protura are of early Devonian origin. This is supported by the phylogenetic placement of Protura as a sister-group to Collembola, and Collembola (the genus *Rhyniella*) being known as Devonian fossils from the Old Red Sandstone of Scotland.

(2) That the evolution in time of the Protura has been a very slow process. This is corroborated by the distribution of *D. tristani* (and of *Brasilentulus*), since the parts of Brazil and West Africa where it is found were divided by the South Atlantic 100 million years ago and it is still the same species on both continents.

(3) That the distribution in space of the Protura has been a very slow process. This, too, is corroborated by the distribution of *D. tristani*, which from its presumed origins has spread only along the Gulf of Guinea in Africa and only to Pernambuco, Amazonas, and Colombia in South America (where it may have been elevated from lowland rainforest to 3500 m in the Andes during the uplift of these mountains in the last 15 million years). The means of dispersal of Protura are limited because these insects are very vulnerable to salt water and susceptible to desiccation.

The place of origin of the Protura is not known — probably in a moist and warm climate, since they are not known in arctic or antarctic countries. But that is not of decisive importance here; it is their presence in New Zealand that matters. My understanding of this I draw from the most inspiring and comprehensively illustrated book *New*

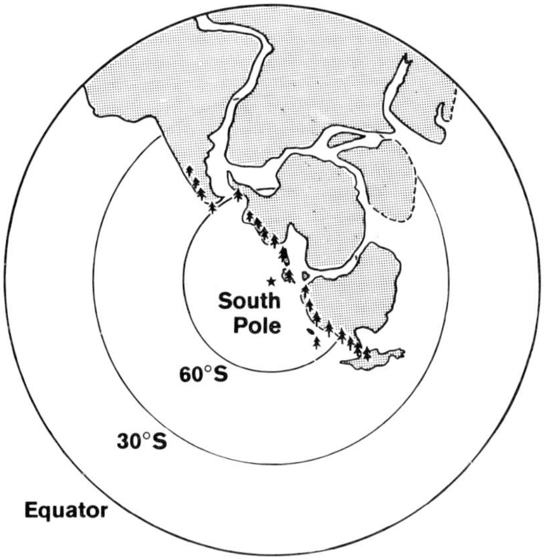

Figure 1. In the early middle Cretaceous (110–90 million years ago) the southern landmasses were closely grouped in cool-temperate latitudes, enabling the southern beeches (*Nothofagus* spp.) and other plants and animals to migrate and achieve a wide distribution. The reconstruction depicted here is a generalised one derived from paleomagnetic studies. *Reproduced, with permission, from Stevens (1980).*

Zealand Adrift, by Graeme Stevens (1980). The association and separation of landmasses used to explain present-day biogeography are based on the supposition of an expanding earth, as set out by Owen (1976, 1981), from a diameter in the early Jurassic of 80% of its present value. This theory has the drawback that nobody knows why the earth should expand, but it helps to understand some problems of distribution.

What is now New Zealand originated at the outskirts of Gondwana. Sediments from Australia and Victoria Land, Antarctica, founded the New Zealand Geosyncline, which collided along an earlier subduction zone with continental sediments from Marie Byrd Land, giving rise to the Rangitata Orogeny in the late Jurassic and early Cretaceous, 140–110 million years ago. New Zealand thus became a big land mass connected to other parts of Gondwana, and could receive plant and animal immigrants from these sources (Figure 1). It is supposed that mosses, kiwis, podocarps, and *Nothofagus* immigrated to New Zealand at this time, and Protura may have come the same way. The position of the subduction zone is still visible; it is characterised by ophiolites, as in the Dun Mountain Ophiolite Belt along the west coast of both main islands. Eighty million years ago, however, in the late Cretaceous, the Tasman Sea between New Zealand and Australia began opening up, isolating New Zealand from the rest of Gondwana. This continued for 20 million years. Fifty-five million years ago the Southern Ocean began opening up between Australia and Antarctica. At this time the South Atlantic had long been in existence, Africa had left Antarctica, and India had started its journey towards Laurasia. The Southern Ocean is still opening up, and New Zealand is placed at the boundary between the Indo-Australian tectonic plate and the Pacific plate. The visible expression of this transform fault is the Alpine Fault.

In Cenozoic time, i.e., for the last 65 million years, New Zealand has sometimes been partly under water, and at other times (e.g., during successive glaciations) has occupied a greater land mass than now. Never was it completely inundated, so survival possibilities for its fauna and flora must always have been present.

A biotic connection across Gondwana from South America via Africa (see Tuxen 1978a), Antarctica, and Australia to New Zealand was thus open at the end of the Jurassic and in the Cretaceous. Such a connection stopped at the beginning of the Tertiary, say 60 million years ago, after which time New Zealand's fauna took its own course — insofar as it took any course. Now, how does this equate with the known distribution within and outside New Zealand of the proturan species known from this country?

If we leave out of consideration the unfortunately indefinable specimen of *Australentulus* (p. 34), fourteen species are here recognised from New Zealand. Of these, eight are new to science, but it is possible to draw conclusions from their systematic relationships. Besides the new species, one more species is endemic to New Zealand. The five species known also from outside New Zealand are:

Eosentomon wygodzinskyi Bonet — known from the Seychelles, Pacific Islands, and Brazil. This is a fine Gondwana distribution.

Proturentomon minimum (Berlese) — known only from Europe; must certainly be an introduced species. In New Zealand found only at Parkes' Farm (NN) on several occasions.

Gracilentulus gracilis (Berlese) — decidedly a European species; probably introduced to the recorded localities in South Africa, Australia, amd New Zealand, though the single New Zealand record is from Little Barrier Island, habitat unknown.

Kenyentulus kenyanus (Condé) — known from Kenya, the Seychelles, southern India, Brazil,

Puerto Rico, and Bermuda. A decidedly Gondwana distribution, with migration to the Caribbean Islands.

Berberentulus nelsoni Tuxen — southern Brazil; in New Zealand, the Kermadec Islands and Pigeon Valley (NN). Decidedly Gondwana. Exactly the same distribution pattern is shown by *Acerentulus kermadecensis* Ramsay & Tuxen, which has clear affinities with the Argentinian species *A. nemoralis* Najt & Vidal.

Leaving apart the two probably introduced species, a Gondwana character is evident in the distribution of all the species known outside New Zealand. Now for thc affinities of the endemic species:

Eosentomon dawsoni Condé and *E. zelandicum* n.sp. — closely related to *E. australicum* Womersley, which is known from Australia and, fairly commonly, from North America (Michigan; E. C. Bernard).

Eosentomon macronyx n.sp., *E. maximum* n.sp., *E. gracile* n.sp. — no close relationship can be established.

Yinentulus paedocephalus n.sp. — no clear relationship.

Amphientulus zelandicus n.sp. and *Tasmanentulus intermedius* n.sp. — distinct relationships to Gondwana species, *Tasmanentulus* embracing two other species from Australia and *Amphientulus* six from Australia (one of them recently found in the Andes; Tuxen 1984), a doubtful one from North Korea, and one from Madagascar.

Thus, if we disregard the two species probably introduced by man, the New Zealand fauna of Protura is distinctly Gondwana in character with connections to the Pacific islands (Solomon Islands, Vanuatu), Australia, southern India, the Seychelles, Madagascar, East Africa, and South America. This means that the species have arrived at the present New Zealand before the opening of the Tasman Sea 80 million years ago, and in all probability in the late Jurassic after the Rangitata Orogeny, when the present New Zealand formed part of a greater land mass connected to the rest of Gondwana.

As to distribution within New Zealand, the records of most species are too scattered to allow of any conclusions. Only two species are sufficiently common to give hints, viz *Amphientulus zelandicus* and *Tasmanentulus intermedius.* Both are known from many localities (see Material Examined, pp. 37 and 40), but whereas *Amphientulus* is found both on the North Island and the northern South Island, *Tasmanentulus* is found only on the South Island, but widely scattered. The two genera are very closely related, and in a few localities have been found together. It might be tempting to suppose the immigration of both species via the South Island, with *Amphientulus* the earliest. Certainly a connection between present distributions and the changing inundation of the islands is more difficult to envisage.

MORPHOLOGY AND DIAGNOSTIC CHARACTERS

Figures 2–13 are scanning electron micrographs that show whole animals and important characters of *Amphientulus zelandicus* and abdominal tergite VIII of *Gracilentulus gracilis.* So as not to spoil the micrographs with labelling I have drawn a specimen in dorsal and ventral aspect (Figures 14 and 15) and marked the characters on them. Other characters are shown in more detail and at greater magnification in the taxonomic drawings (Figures 22–165).

The three tagmata — head, thorax, and abdomen — are easily distinguished. On the head the following characters are diagnostically important.

The pseudoculi (*ps*) are probably homologous with the postantennal organ in Collembola. They may be round or oval, with or without structures, or with a small or bigger "lever" posteriorly. The length of the head behind the rostral setae divided by the length of the pseudoculus without the lever is termed *PR*.

The mandibles (*mdb*) may be broad apically with from two to four minute teeth or taper into a pointed tip.

The maxillary palps (*mxp*) are three-segmented and end in a tuft of setae. The last segment but one carries two sensillae,* the shape of which may be important.

The canal of the maxillary gland (*mxgl*) is a single tube in Eosentomoidea but is varied and systematically very important in Acerentomoidea. Originating from the gland as a variously shaped bulb, it is tubular proximally, and has a widening calyx before the final duct. The maxillary gland canal, originally called "filamento di sostegno", is of ectodermal origin and easily seen in cleared specimens. The length of its proximal part (i.e., proximal to the calyx) may be compared with the proximal branch of the fulcrum (*fu*), which is the tentorium-like internal skeleton of the entognathous head.

---

*The author's usage of "*sensilla, -ae*" follows his own practice of many years, and that of other workers on Protura, although not strictly correct in terms of the word's Latin origins; a fact which he has many times acknowledged in print. As Dr G. Kuschel points out, the word is a diminutive of *sensus*, and it should take the form "*sensuillus*" — though "*sensillus*" comes more readily to the tongue and to the pen. (–Editor).

Figures 2–13. Scanning electron micrographs of *Amphientulus zelandicus* (2–11) and *Gracilentulus gracilis* (12 and 13): 2, lateral view; 3, ventral view; 4, dorsal view; 5, head, ventral view; 6, head, dorsal view; 7, labial palps; 8, foretarsus, lateral view; 9, pseudoculus; 10–13, tergite VIII and striate band. Scale lines: 200 µm (2–4); 40 µm (6); 20 µm (5, 8); 10 µm (7, 11, 13); 4 µm (9, 10, 12). *SEMs by Bent W. Rasmussen.*

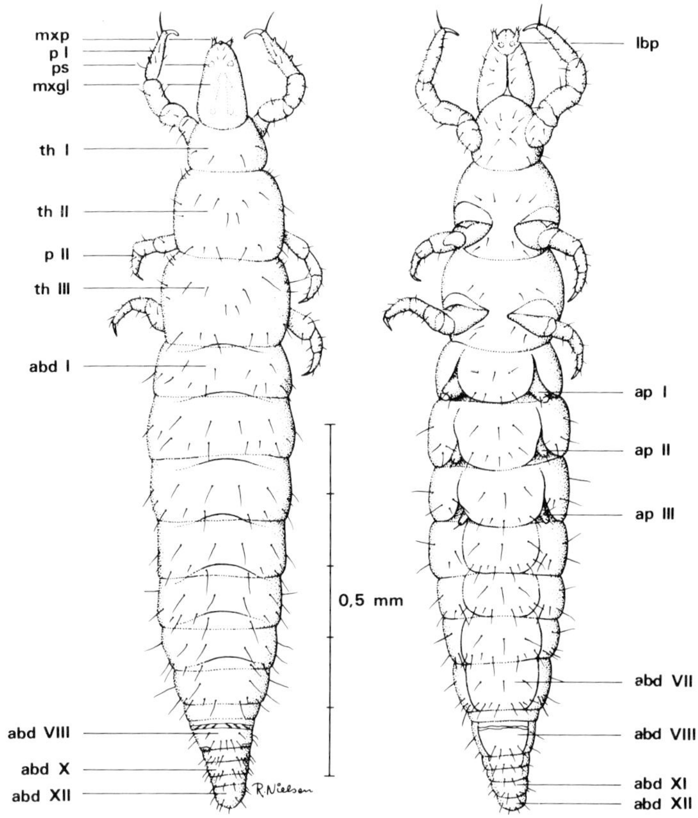

(14) (15)

Figures 14 and 15. *Amphientulus zelandicus* in dorsal (14) and ventral (15) view. Scale line 5 × 0.1 mm. Key: abd, abdominal segments; ap, abdominal legs; lbp, labial palp; mxgl, maxillary gland; mxp, maxillary palp; p, thoracic legs; ps, pseudoculus; th, thoracic segments.

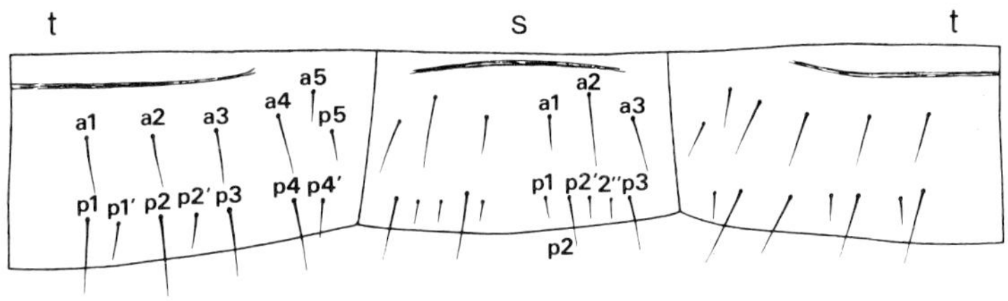

(16)

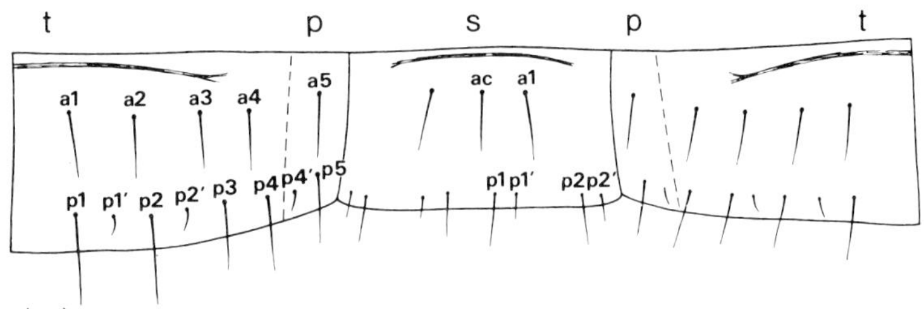

(17)

Figures 16 and 17. Chaetotaxy of abdominal segment IV in an eosentomid (16) and an acerentomid (17); modified from Tuxen (1949). Key: p, pleuron; s, sternum; t, tergum; all other characters are setal notation.

The labial palps (*lbp*) are small, usually one-segmented, ending in a tuft of setae like the maxillary palp or reduced, with three or four setae and an often sausage-like sensilla.

On the thorax the most important characters, besides the chaetotaxy, are associated with the foretarsus (see, e.g., Figures 22 and 23). This is generally long and slender, and terminates in a claw attached to a small pretarsus on which a ventral empodium (*emp*) and a dorsal sensilla (*s*) are attached. The length of the foretarsus proper (excluding pretarsus and claw) divided by the length of the claw plus pretarsus is called *TR* (tarsus ratio). The length of the empodium divided by the length of the claw proper is termed *EU*. The tarsus carries a number of sensillae usually with rounded tips, and a number of setae (with pointed tips) which are designated with letters and numbers and are of extreme systematic value.

Sensillae. There are three dorsally, *t1*, *t2*, and *t3*. On the exterior side there are at most seven, labelled *a–g*, in Acerentomoidea but an additional *f2*, *x*, *y*, and *z* in Eosentomoidea. On the interior side Acerentomoidea have at most three sensillae, *a'*, *b'*, and *c'*; and Eosentomoidea have an additional *b'2*. Sensilla *t1* may be club-shaped (claviform) or have a ball-shaped head (baculiform).

Setae. These are arranged in a certain pattern on the four "sides" of the foretarsus. Dorsally there is a zigzag line of seven setae in Acerentomoidea, *α1–7*; and in Eosentomoidea an additional *α3'* in line with *α3*. Ventrally another zigzag line of seven setae, *β1–7*, is found in Acerentomoidea; in Eosentomoidea the line continues up to *β9*. Seta *β1* is always very small. On the exterior side five setae, *γ1–5*, are found more or less in a zigzag line in both suborders. Finally, the four proximal setae on the interior side, *δ1–4*, are arranged in an oblique line beginning dorsally, while *δ5* and *δ6* are closer to the midline. In Eosentomoidea there are an additional *δ3'* and *δ4'* dorsal to *δ3* and *δ4*. This sounds complicated, but in reality is usually easily recognised, and is important for the identification of the sensillae.

From the designation of the sensillae and setae it might seem as if the Eosentomoidea are more specialised than the Acerentomoidea, but I do not think they are. The reason is historical: the designations were at first applied to acerentomid species (by Condé and myself) and later transferred to eosentomids.

The position of *t1* is important, especially in Eosentomoidea; it is called *BS*, namely the distance to tarsal base divided by the distance to tarsal apex (without claw).

The middle and hind legs are of systematic interest only in connection with their claws: these are knife-shaped in Eosentomoidea, but in Acerentomoidea more or less boat-shaped. In Eosentomoidea the length of the empodium relative to the claw (*EU*) is of importance.

In Eosentomoidea the mesothorax and metathorax carry one spiracle each on either side, but the tracheae of each do not meet.

The abdomen consists of twelve segments, or eleven plus a telson. The anus opens on the telson, and the genital organs open between this and the eleventh segment. The following characters are of importance.

The first three abdominal segments each carry a pair of rudimentary legs (e.g., Figures 82–84). The first pair is always two-segmented, ending in a terminal vesicle; in Eosentomoidea it always has five setae, in Acerentomoidea never more than four. The second and third pair may be like the first (Eosentomoidea) or a rudimentary bud without a vesicle and with one, two, or three setae, one of which is always big, subapical, and lateral and the two others smaller, and lateral or medial.

The eighth abdominal segment carries posteriorly on either side the opening of the big abdominal glands. This opening is always covered by a "lid", which in Eosentomoidea is undifferentiated but in Acerentomidae is diverse, with or without teeth differing in number and length; it is often called comb VIII.

In Acerentomoidea a so-called "striate band" is present proximally on the eighth abdominal segment; it is most differentiated dorsally. In Protentomidae it is only a transverse line, but in Acerentomidae there are two transverse lines with or without structures in between. These structures may take the form of a cuticular grate of stakes (hence the name), but even where these are missing one may imagine their presence because corresponding subdermal structures are visible, probably glandular tubules. Even where the stakes are missing a cuticular structure may be seen (see Figures 10 and 11).

The external genitalia differ from those of all other insects. In both sexes they open ventrally between the eleventh segment and telson (the gonotreme). In both sexes there are a pair of basal arms and two styli. The male arms, or periphallus, carry setae but the female arms, or perigynium, never do. The gonopore is on the styli, and thus double, in the male but between the styli, and thus single, in the female. The male genitalia are generally alike in all species, though differing between Eosentomoidea and Acerentomoidea, whereas the genitalia show distinct specific differences in the female. In Eosentomoidea the styli most often carry a curious structure, called the processus sternalis (plural: processūs sternales), which is of characteristic shape between species and thus of great taxonomic importance. In Acerentomoidea there is no such structure; each of the styli ends in an acrostylus, the shape of which may also be of taxonomic value (see Figures 29 and 117; also Tuxen 1970, pp. 21–24).

Chaetotaxy. The position, size, and number of setae dorsally and ventrally on the thorax and abdomen are of the utmost taxonomic value. They were designated by me in 1949 and tabulated in 1964. It is true that some variability in number of setae may be found within the species (in 1961 I examined the intraspecific variability in 476 specimens of one species, *Acerentomon gallicum* Ionescu), but used with care, chaetotaxy is an important specifically distinctive character. In principle the chaetotaxy is as follows.

The pronotum almost always has four setae in a single row.

The mesonotum and metanotum each have an anterior row of equally long setae, a median pair of setae, and a posterior row with principal and accessory setae as on the abdominal segments.

Abdominal terga (Figures 16 and 17): the setae are arranged in two transverse rows on each side. The anterior row is designated *a1–5*, since the setae are generally of equal length. In the posterior row the setae are arranged into long, strong principal setae (*p1–5*) and much thinner accessory setae (*p1′–4′*). The accessory setae may be longer or shorter than the principal ones, but are always more gracile, and may be placed just anterior to the hind border or in an excavation of the hind border. (They are designated as *1a*, etc. instead of *1′* by Imadaté and others.) On tergites IX–XI only one row is present on each side.

Thoracic sterna: their chaetotaxy is not used in taxonomy, being uniform throughout the Protura. Abdominal sterna (Figures 16 and 17): the setae on sternites I–VII are arranged into two rows, the posterior row with principal and accessory setae; but on sternites IX–XI there is only one row. The eleventh sternite always carries four fewer setae in the maturus junior than in the adult.

For descriptive purposes chaetotaxy can be tabulated into big schemata with every seta in every stage stated, or expressed according to "Tuxen's scheme", with the numbers of setae in the anterior and posterior rows given respectively above and below a line, and then notes about important occurrences or absences of single setae. The latter scheme is used in this contribution.

LIFE CYCLE AND BIOLOGY

As already mentioned, Protura are the only insects with anamorphosis, meaning that they add body segments during development. Berlese knew this, but thought that only one segment was added from each stage to the next. In 1949 I proved this to be incorrect, and described the true development. The names I then attached to the different stages may not have been very happily chosen, but they have been used ever since.

Protura hatch as a prelarva (proved without doubt by Bernard (1976)), which probably is immobile. It has only nine abdominal segments and relatively undeveloped mouthparts, and often the thoracic and abdominal legs are poorly developed (Figures 18–21). The prelarva develops into a larva I with nine abdominal segments and fully developed mouthparts and legs. From this stage it develops into a larva II with ten abdominal segments (the new stage is often discernible in the cuticle of the preceding one if they are cleared with, e.g., lactic acid or other agents). This develops into a maturus junior (the name given by Womersley) with twelve abdominal segments, often a smaller number of setae, and no external genitalia. From this, finally, develops the adult — though often a preimaginal male stage with incomplete external genitalia intervenes.

Development takes place entirely in the soil or in decaying wood, where Protura may occasionally be found (Tuxen 1931). In my 1931 paper I supposed a relationship between the pH of the soil and the presence of Protura; in acid soils their abundance was far greater than in alkaline ones, where often they could not be found at all. Later papers by other authors have to some degree corroborated this result, but more thorough investigations have now been made. In my papers of 1931 and 1949 I supposed the mineral soil horizons to be more or less free of Protura, but investigations in Brazil have shown that this is not always true. It may be, as Yin (1981) showed, that different genera descend to different depths in the soil, and thus are found most abundantly at different levels. Also, their yearly cycles may be different; in 1949 I found a Danish *Eosentomon* species that was what I called euryplastic, with young stages all the year round, whereas an acerentomid species was stenoplastic, with a distinct breeding season.

As to the diet of Protura, Sturm (1959) showed that two European species, an *Eosentomon* and an *Acerentomon*, where feeding on mycorrhiza on beech. In Brazil I used this experience to help me decide under which trees to look for Protura, but no further investigations have been made.

As to the ecology of New Zealand Protura, nothing can be said. The localities are always given as

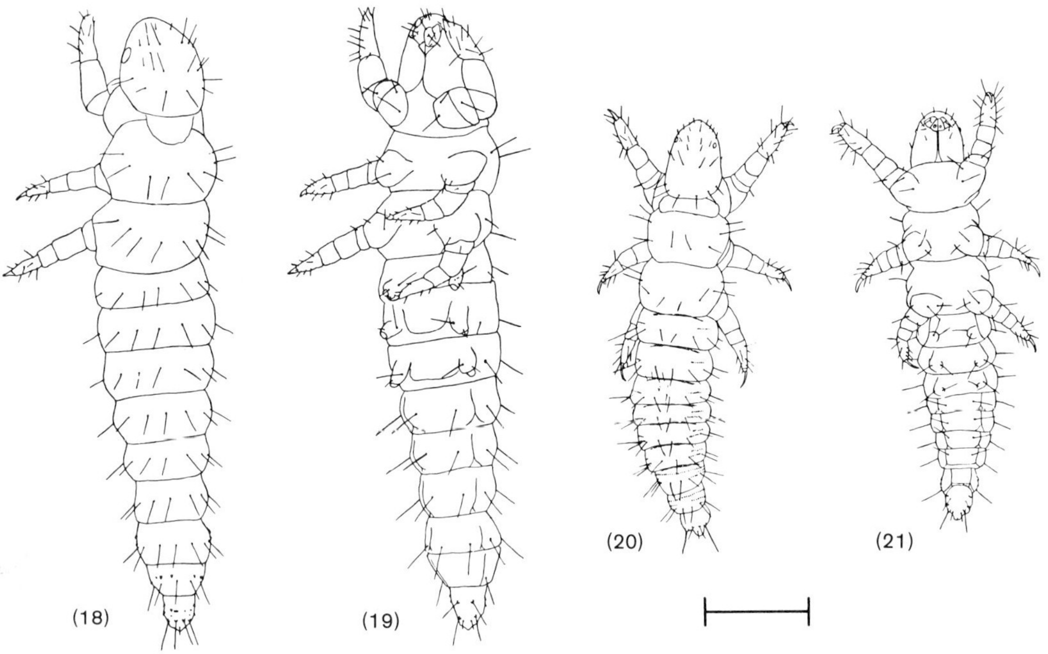

Figures 18–21. Prelarva of an eosentomid (18, 19) and of an acerentomid (20, 21), in dorsal and ventral view (scale line 100 μm); from Tuxen (1949).

"moss and litter" under trees and bushes of many species, though often *Nothofagus*. This may have a connection with mycorrhiza, but may also be a collection artefact reflecting the abundance of this tree genus. Nothing is even stated about the depth range of the samples [Usually shallow. -Editor] or the composition of the soil. A wide field is open to New Zealand soil biologists.

COLLECTING, PREPARATION, AND STUDY

Being very small and white or colourless animals with slow movements, the Protura are difficult to catch sight of with the naked eye. I know (and have seen myself) that some Chinese proturologists can; but I have always collected them with the aid of Berlese funnels or Tullgren funnels. The sample is placed on a wire screen at the top of the funnel, and is dried by means of hot water surrounding it (Berlese) or by light from above (Tullgren). The animals leave the soil and are collected below; the Protura are then sorted out. This treatment may be used for quantitative analysis if the soil sample is taken with a corer of known size and the sample is divided into subunits from different depths.

Under the stereomicroscope it may be possible to distinguish between the genera or even species of fresh material, but most often it is not, and so the specimens must be cleared and prepared as slide mounts. The solution used for clearing may be Faure's fluid or Hoyer's medium or other agents, but all are derived from Berlese's medium and contain chloral hydrate. For clearing, the slides must be put immediately into an oven at about 55°C and stay there for some days, preferably a week, so that they are quite dry. They may be sealed with, e.g., epoxy resin or fingernail polish. The slides are studied under a compound microscope, and oil-immersion objectives of ×100 magnification may be necessary. Phase-contrast objectives may be advantageous when there is any doubt.

For identification, Tuxen (1964) is the earliest monograph, but identification keys have been published more recently for the south-east Asian Protura (Imadaté 1965), the Australian ones (Tuxen 1967), the Japanese (Imadaté 1974), the European (Nosek 1973), and the Brazilian (Tuxen 1976), to mention some faunas of relevance to the New Zealand situation.

The material examined for this study belongs to the following institutions:

CMNZ Canterbury Museum, Christchurch
LCNZ Department of Entomology, Lincoln College, Canterbury
NMNZ National Museum of New Zealand, Wellington
NZAC New Zealand Arthropod Collection, Entomology Division, DSIR, Auckland
and now partly to
ZMKD Zoologisk Museum, København, Denmark.

All measurements, drawings, and descriptions are based on New Zealand material; this includes species previously known from elsewhere. All drawings were prepared by myself.

Abbreviations, other than seta and sensilla designations, used in the text and figures:

abd	abdominal segments
ap	abdominal legs
BS	position on tarsus of sensilla *t1* (distance to base / distance to apex)
emp	empodium
EU	empodium / claw length ratio
fu	fulcrum
lbp	labial palp
mdb	mandible
mxgl	maxillary gland
mxp	maxillary palp
p	thoracic legs
pl	pleuron
PR	head length / pseudoculus length ratio
ps	pseudoculus
s	pretarsal sensilla
st	sternum
t	tergum
th	thoracic segments
TR	tarsus ratio (tarsus proper / claw + pretarsus)

KEY TO FAMILIES OF PROTURA

01 Spiracles present on mesothorax and metathorax (exception: *Anelientomon*); claw of middle and hind legs knife-shaped; lacinia of maxilla hook-shaped; pseudoculus a simple elevation; abdominal legs 2-segmented, with 5 setae; no striate band on abdominal tergite VIII; female squama genitalis most often with a processus sternalis
... **EOSENTOMOIDEA, EOSENTOMIDAE**

— Spiracles absent; claw of middle and hind legs boat-shaped; lacinia of maxilla pointed; pseudoculus with a rim and other structural differentiations; abdominal legs 2-segmented, with at most 4 setae, or (legs II and III only) 1-segmented; striate band on abdominal tergite VIII present at least as a serrate line; female squama genitalis with acrostyli but no processus sternalis ... **ACERENTOMOIDEA** .. 2

02(01) Only abdominal leg I 2-segmented, with 4 setae; striate band on abdominal tergite VIII a double line with or without striate structures in between (see Tuxen 1981, 1983)
... **ACERENTOMIDAE**

— Abdominal legs I, I and II, or I–III 2-segmented with at most 4 setae, the remainder 1-segmented with 1–3 setae; striate band present only as a serrate line ... **PROTENTOMIDAE**

KEY TO PROTURA KNOWN FROM NEW ZEALAND

(Because so few species are known, the characters are arranged to lead directly to species.)

01 Tracheal system present; all abdominal legs with a terminal vesicle and 5 setae
... **EOSENTOMOIDEA, EOSENTOMIDAE** .. 2

— Tracheal system absent; only the 1st abdominal leg, or 1st and 2nd, with a terminal vesicle and at most 4 setae
... **ACERENTOMOIDEA** .. 7

02(01) Abdominal sterna IX and X each with 6 setae; *p1″* and *p2* on tergite VIII displaced anteriorly
... (p. 27) .. ***Eosentomon gracile***

— Sterna IX and X each with 4 setae; *p1″* and *p2* not displaced ... 3

03(02) Tergum XI with 4 setae; tergum X with 2 setae
... (p. 16) .. ***Eosentomon dawsoni***

— Tergum XI with 8 setae; tergum X with 8 setae or none ... 4

04(03) Tergum X without setae; tergum VII with 2 anterior setae (*a5*)
... (p. 18) .. ***Eosentomon wygodzinskyi***

— Tergum X with 8 setae; tergum VII with 4 anterior setae (*a4* and *a5*) ... 5

05(04) *p4′* present on terga II and III; caput processus on female squama genitalis semicircular
... (p. 20) .. ***Eosentomon macronyx***

— *p4'* absent on terga II and III; caput processus sharply bent against median line ... 6

06(05) Very large species — length of foretarsus 105 μm; *p1'* on tergum VII close to *p2* ... (p. 25) .. ***Eosentomon maximum***

— Smaller species — length of foretarsus around 80 μm; *p1'* on tergum VII inserted on hind border ... (p. 23) .. ***Eosentomon zelandicum***

07(01) 1st and 2nd abdominal legs with a terminal vesicle ... (p. 30) .. PROTENTOMIDAE, ***Proturentomon minimum***

— Only the 1st abdominal leg with a terminal vesicle ... ACERENTOMIDAE .. 8

08(07) Striate band with a distinct and regular cuticular grate of stakes (Fig. 12, 13) ... 9

— Striate band reduced, the striae at most indistinctly present, as if below the cuticle (Fig. 10, 11) ... 11

09(08) 2nd and 3rd abdominal legs each with 3 setae ... 10

— 2nd and 3rd abdominal legs each with only 2 setae ... (p. 34) .. ***Gracilentulus gracilis***

10(09) Labial palp well developed, with a tuft ... (p. 32) .. ***Acerentulus kermadecensis***

— Labial palp reduced to a knob, with a sensilla and 3 setae ... (p. 34) .. ***Australentulus*** sp. indet.

11(08) Canal of maxillary gland with 2 or 3 dilatations proximal to calyx ... (p. 42) .. ***Kenyentulus kenyanus***

— Canal of maxillary gland simple ... 12

12(11) 1st and 2nd abdominal legs each with a subapical seta and a fairly large lateral apical seta ... 13

— 1st and 2nd abdominal legs each with a subapical seta and a small, delicate, median apical seta ... (p. 46) .. ***Berberentulus nelsoni***

13(12) Sensilla *t1* of foretarsus baculiform ... (p. 42) .. ***Yinentulus paedocephalus***

— Sensilla *t1* of foretarsus claviform ... 14

14(13) Sternite VIII without posterior setae; sensilla *b* on foretarsus as long as *c* ... (p. 37) .. ***Amphientulus zelandicus***

— Sternite VIII with 2 posterior setae; sensilla *b* on foretarsus only half as long as *c* ... (p. 40) .. ***Tasmanentulus intermedius***

DESCRIPTIONS

Suborder EOSENTOMOIDEA

Tracheal system present or reduced. Abdominal legs each with a terminal vesicle and 5 setae. Claw of middle and hind legs knife-shaped. Lacinia of maxilla hook-shaped. Pseudoculus a simple elevation. Abdominal tergum VIII without a striate band or serrate line. Female squama genitalis most often with a processus sternalis but without acrostyli. Sternum XII with 12 setae.

Monobasic.

Family EOSENTOMIDAE

Diagnosis as for suborder Eosentomoidae.

Genus *Eosentomon* Berlese

Berlese, 1908: 18. Type species *Eosentomon transitorium* Berlese, 1908, by original designation.

Mesothorax and metathorax with spiracles. Mandible rather broad and stout, ending usually in 3 teeth, and striate in its distal part. Sensillae *e* and *g* on foretarsus spatulate. Terga X and XI without a huge spine.

Eosentomon dawsoni Condé

Figures 22–29

Condé, 1952: 163–165. Tuxen, 1964: 123–124.

Length of body (extended) 800 μm. Length of foretarsus without claw 79 μm.

HEAD. Mouthparts of the common eosentomid type. Mandibles ending in 3 small teeth. Labral setae present. Clypeal apodeme with a well developed anterior connection (Figure 24). Pseudoculus almost circular, with a central globule (Figures 25 and 26); *PR* = 12–13.

THORAX. Foretarsus (Figures 22 and 23) with all sensillae present; *e* and *g* spatulate; *f2* almost as long as *f1*, longer than *t3*; *b' 1* present; *c'* not especially thick; *a* almost reaching $\gamma 2$, *c* just surpassing $\gamma 3$; *t1* about midway between $\alpha 3$ and $\alpha 3'$ (*BS* = 1.2); empodium as long as claw (*EU* = 1.0); *TR* = 4.2. Empodium of middle and hind tarsi one-fifth as long as claw; spine on hind tarsus distinct (Figure 27).

Chaetotaxy of metanotum: *p1'* long, behind the line *p1–2*.

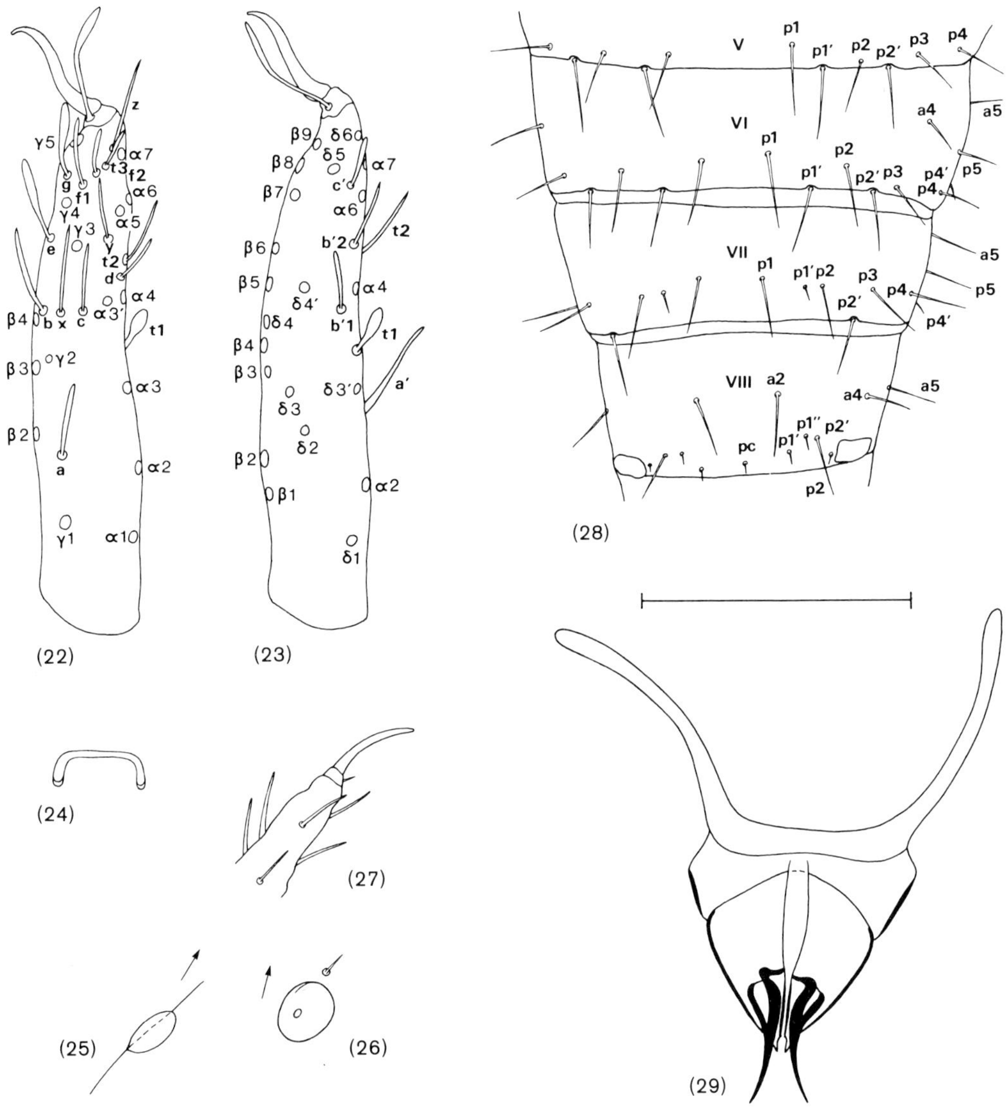

Figures 22–29. *Eosentomon dawsoni*: 22, 23, foretarsus, anterior and interior view; 24, clypeal apodeme; 25, 26, pseudoculus; 27, hindtarsus; 28, tergal chaetotaxy of abdominal segments V–VIII; 29, squama genitalis of female. Scale line: 100 μm (28); 40 μm (24–27); 26 μm (29).

ABDOMEN. Chaetotaxy of segments as in Table 1 and Figure 28.

On terga V and VI only *a4* and *a5* present, and on tergum VII only *a5*; on tergum X only the 2 laterals, and on tergum XI 4 laterals; *p4′* present on terga II–VII; *p1′* long, surpassing *p1* on terga II–VI, encroaching on hind border, but very short and on a line with *p1–2* on tergum VII; on tergum VIII *p1″* and *p2* not displaced anteriorly.

Squama genitalis of female (Figure 29) with caput processus shaped like a duck's head, squarely bent against median line; corpus processus with an incision; very weak proximolateral sclerotisations; filum processus fairly long.

Table 1 Abdominal chaetotaxy of *Eosentomon dawsoni*

	I	II,III	IV	V,VI	VII	VIII	IX	X	XI	Telson
Tergum	4/10	10/16	10/16	4/16	2/16	6/9	8	2	4	9
Sternum	4/4	4/4	10/10	10/10	10/10	7	4	4	8	12

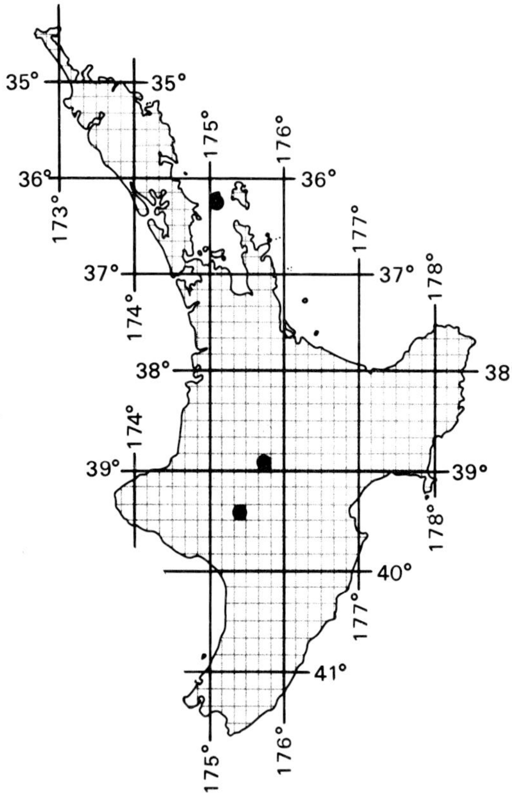

Type data. **Lectotype** female, New Zealand, CL, Little Barrier Island, 20 November 1949, E. W. Dawson (CMNZ).

Material examined. Lectotype, plus 15 non-type examples (7 females, 4 maturi juniores, 3 larvae II; NMNZ, NZAC, ZMKD).

CL (Little Barrier I.), TO / —.

Collected in January, October, and November.

Taken from wood mould under *Prumnopitys taxifolia* (matai; formerly *Podocarpus spicatus*) and from leaf litter of *Dacrydium cupressinum* and *Nothofagus fusca.*

Remarks. The above description is based on one of two females from Te Ponanga Bush, Tokaanu (TO). It deviates from the original description (Condé 1952) and the redescription by Tuxen (1964) in the length of the body and foretarsus and in *TR*. The foretarsi are missing in both the lectotype and the maturus junior found with it; the *TR* value given by Tuxen was based on Condé's drawing, not the specimen. The lectotype is now split up into several pieces. However, measurement of abdominal segments I–VI gives 420 µm (350 µm on the specimens from Te Ponanga), and by extrapolation the length of the lectotype should thus be about 950 µm.

This species is known only from New Zealand.

The name was given in honour of the collector of the lectotype, Mr E. W. Dawson.

Eosentomon wygodzinskyi Bonet

Figures 30–36

Bonet, 1950: 122–123. Tuxen, 1964: 137–138; –1976: 435; –1977: 301–302; –1978b: 255.

solare Tuxen & Imadaté, 1975: 356–358. Synonymised by Tuxen (1977b, p. 301).

Length of body (extended) 1100 µm. Length of foretarsus without claw 97 µm.

HEAD. Mouthparts of the common eosentomid type. Mandibles ending in 3 small teeth. Labral setae present. Clypeal apodeme with a well developed anterior connection (Figure 32). Pseudoculus circular, with a central globule (Figure 33); *PR* = 14–16, rarely 12–13.

THORAX. Foretarsus (Figures 30 and 31) with all sensillae present; *e* and *g* spatulate; *f2* half as long as *f1*, broadened apically; *b′ 1* present, *c′* normal; *a* reaching *γ2*; *c* far surpassing *γ3*; *t1* midway between *α3* and *α3′* (*BS* = 1.1–1.2); empodium shorter than claw (*EU* = 0.9); *TR* = 5.0 (4.6–5.3). Empodium of middle and hind tarsi less than one-fifth as long as claw; spine and hind tarsus distinct (Figure 34).

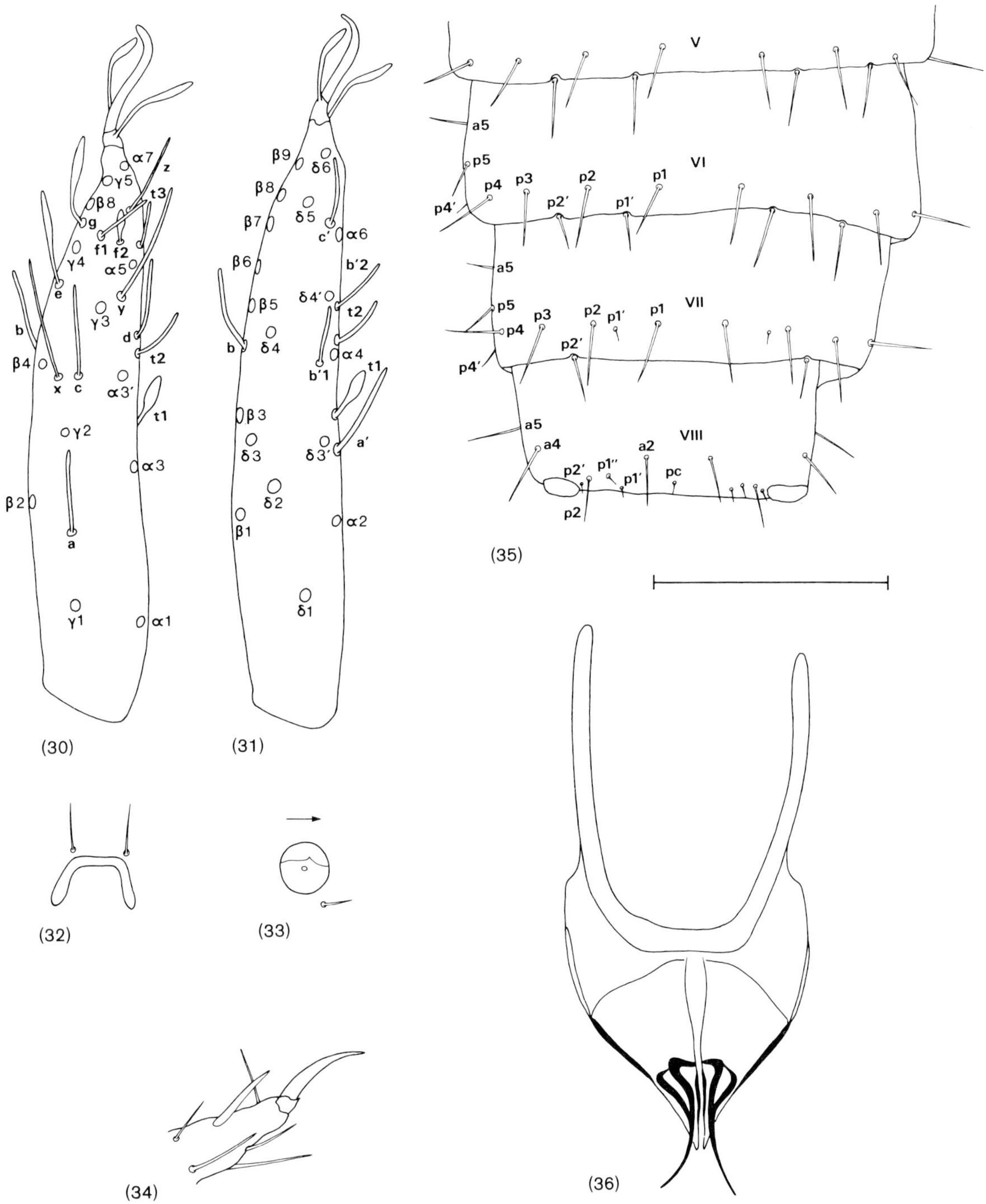

Figures 30–36. *Eosentomon wygodzinskyi*: 30, 31, foretarsus, anterior and interior view; 32, clypeal apodeme; 33, pseudoculus; 34, hindtarsus; 35, tergal chaetotaxy of abdominal segments V–VIII; 36, squama genitalis of female. Scale line: 100 μm (35); 40 μm (30–34); 26 μm (36).

Inset 4

Table 2 Abdominal chaetotaxy of *Eosentomon wygodzinskyi*

	I	II,III	IV	V,VI	VII	VIII	IX	X	XI	Telson
Tergum	4/10	10/16	10/16	4/16	2/16	6/9	8	0	8	9
Sternum	4/4	6/4	6/10	6/10	6/10	7	4	4	8	12

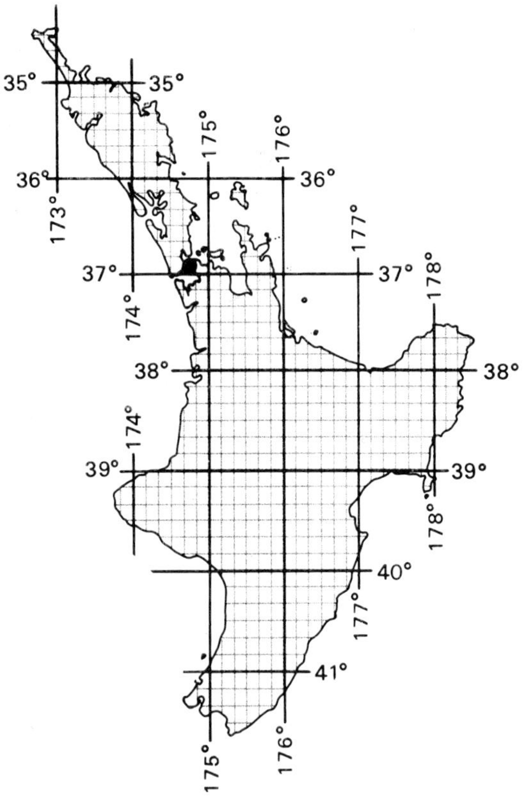

Chaetotaxy of metanotum: *p1′* long, behind the line *p1–2*.

ABDOMEN. Chaetotaxy of segments as in Table 2 and Figure 35.

On tergum V only *a4* and *a5* present, and on tergum VI and VII only *a5*; on tergum X all setae missing; on tergum XI all setae present, the 4 median ones microchaetae; *p4′* present on terga II–VII; *p1′* long, surpassing *p1* on terga II–VI, encroaching on hind border, but very short and on a line with *p1–2* on tergum VII; on tergum VIII *p1″* not displaced anteriorly.

Squama genitalis of female (Figure 36) with caput processus shaped like a duck's head, squarely bent against median line; corpus processus with an incision; filum processus long, slender.

Type data. Holotype female, Brazil, Rio de Janeiro, Itaguai, June 1945, P. Wygodzinsky (Bonet Collection; now in ZMKD).

Material examined. Holotype, plus 16 non-type examples (4 males, 11 females, 1 larva II; NZAC, ZMKD).

AK / —.

Collected in April, May, October, and December.

Taken from soil under the moss *Ptychomnion aciculare* and around roots of *Metrosideros excelsa*.

Remarks. *E. wygodzinskyi* is variable in the chaetotaxy of tergites VI and X (see Tuxen 1977, p. 302). All specimens from New Zealand agree with eight examples from Aneityum and Erromanga, Vanuatu (formerly New Hebrides), in having two *a*-setae on tergum VI and none on tergum X, differing in this respect from all other specimens. Also, the size of the pseudoculus varies in the New Zealand specimens, from *PR* = 12 to *PR* = 16.

This species has been recorded also from Brazil, the Solomon Islands and Bismarck Archipelago (under the name *solare*), Vanuatu, and the Seychelles.

The name was given in honour of the collector of the holotype, the well known specialist in Thysanura Dr P. Wygodzinsky.

Eosentomon macronyx new species

Figures 37–45

Length of body 1200 μm. Length of foretarsus without claw 94 μm.

HEAD. Mouthparts of the common eosentomid type. Mandibles ending in 3 small teeth. Labral setae present. Clypeal apodeme with a well developed anterior connection and very large bulbs (Figure 39). Pseudoculus circular, with a central globule (Figure 40); *PR* = 11.

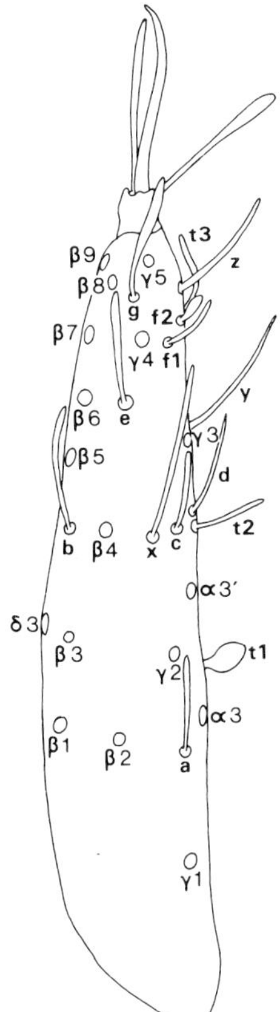

(37)

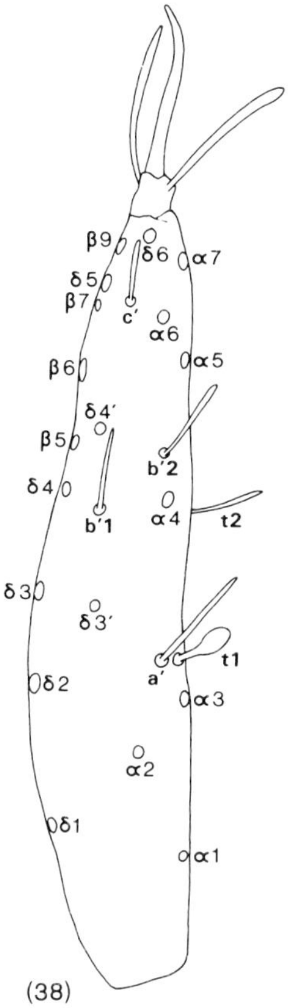

(38)

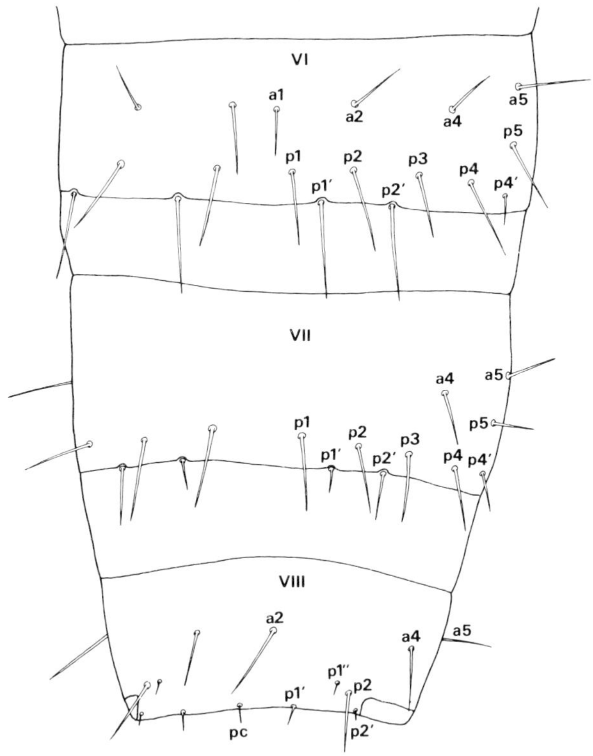

(42)

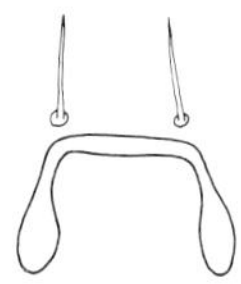

(39)

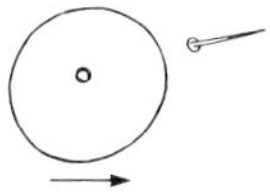

(40)

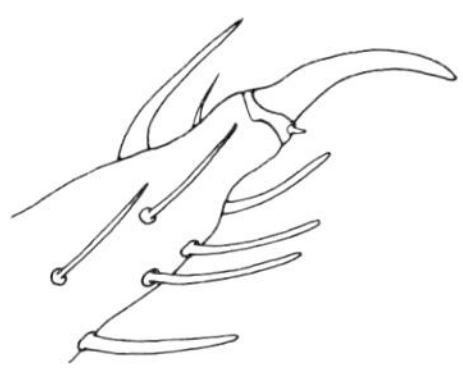

(41)

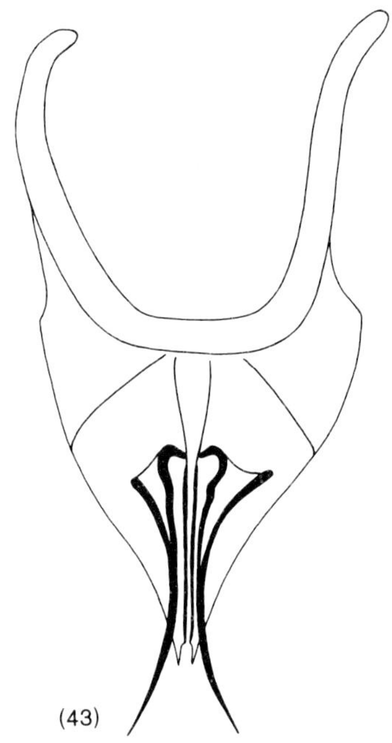

(43)

Figures 37–43. *Eosentomon macronyx*: 37, 38: foretarsus, exterior and interior view; 39, clypeal apodeme; 40, pseudoculus; 41, hindtarsus; 42, tergal chaetotaxy of abdominal segments VI–VIII; 43, squama genitalis of female. Scale line: 100 µm (42); 40 µm (37–41); 26 µm (43).

Table 3 Abdominal chaetotaxy of *Eosentomon macronyx*

	I	II,III	IV–VI	VII	VIII	IX,X	XI	Telson
Tergum	4/10	10/16	8/16	4/16	6/9	8	8	9
Sternum	4/4	6/4	6/10	6/10	7	4	8	12

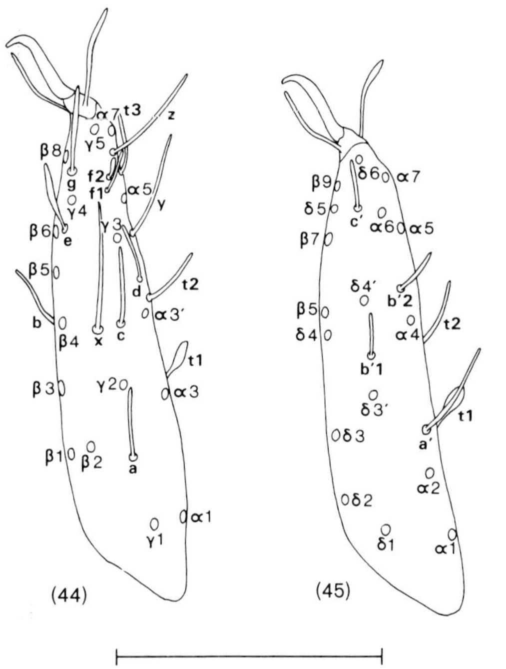

Figures 44 and 45. *Eosentomon macronyx*, foretarsus in exterior and interior view; cf. Figures 37, 38, and text p. 23. Scale line 40 µm.

THORAX. Foretarsus (Figures 37, 38, 44, and 45) with all sensillae present; *e* and *g* spatulate but slender; *f2* slightly more than half as long as *f1*; *b′ 1* present; *c′* normal; *a* reaching *γ2*; *c* reaching *γ3*; *t1* closer to *α3* than to *α3′* (*BS* = 0.8); empodium as long as claw (*EU* = 1.0); claw relatively large (*TR* = 4.0). Empodium of middle and hind tarsi less than one-fifth as long as claw; spine and hind tarsus distinct (Figure 41).

Chaetotaxy of metanotum: *p1′* long, behind the line *p1–2*.

ABDOMEN. Chaetotaxy of segments as in Table 3 and Figure 42.

On terga IV–VI *a3* missing, and in some specimens also on tergum III or even II; on tergum VII only *a4* and *a5* present; on tergum X all setae present; *p4′* present on terga II–VII; *p1′* long, surpassing *p1* on terga II–VI, encroaching on hind border, but short and encroaching on hind border on tergum VII; on tergum VIII *p1″* and *p2* not displaced anteriorly.

Squama genitalis of female (Figure 43) with caput processus semicircular, bent against median line; corpus processus slender, with an incision; stylus long, pointed; filum processus fairly long, slender.

Type data. Holotype female, New Zealand, WD, Simonin Pass, western Olivine Range, 1077 m, *Nothofagus* forest edge, 23 January 1975, G. W. Ramsay (NZAC).

Material examined. Holotype, plus 25 non-type examples (10 females, 8 males, 2 maturi juniores, 2 larvae II, 3 larvae I; NMNZ, NZAC, ZMKD).

ND, AK, TO / WD.

Collected in January, May, and October–December (pre-adult stages: maturus junior — October; larvae II and I — December).

Taken from soil around roots of *Metrosideros* spp. and *Leptospermum scoparium*, soil under the moss *Ptychomnion aciculare*, and leaf litter of *Dacrydium cupressinum*.

Remarks. *E. macronyx* may vary in the chaetotaxy of terga II and III, and in size — two specimens from Lynfield (AK) are smaller. In two specimens from Waipoua (ND) *b′ 1* is missing. This may be a specific character, but owing to the paucity of specimens, in which one foreleg or even both may be missing, nothing can be decided at present. Also, the relative length of the foretarsal claw may vary (*TR* = 4.0–4.8), so the meaning of the name (see below) holds true only for the holotype. The abdominal chaetotaxy, the position of *t1* on the foretarsus (close to *α3*), and the shape of the caput

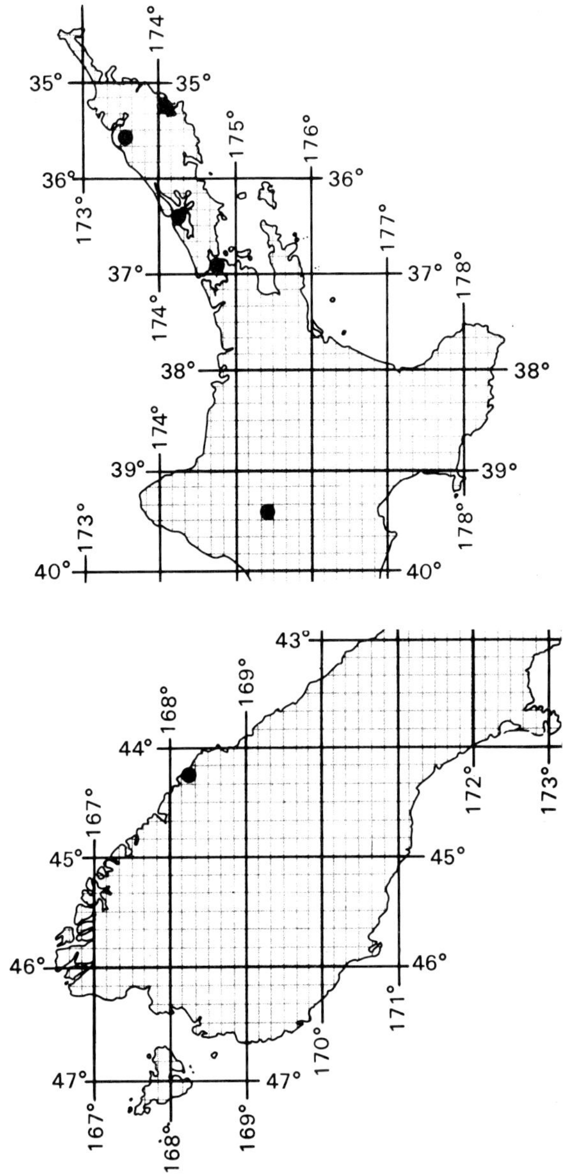

processus in the female squama also characterise this species. Sensillae *d* and *t2* relative to *b–c* may differ, and I have depicted the foretarsus of two extremes (the holotype, Figures 37 and 38, and another specimen, Figures 44 and 45). However, I have found no conformity in these variations, and have included them all under *macronyx*. Further material may show whether I am right to do so.

This species is known only from New Zealand.

The name refers to the relatively large size of the foretarsal claw.

Eosentomon zelandicum new species

Figures 46–52

Length of body 870 μm. Length of foretarsus without claw 80 μm.

HEAD. Mouthparts of the common eosentomid type. Mandibles ending in 3 small teeth. Labral setae present. Clypeal apodeme with a well developed anterior connection (Figure 48). Pseudoculus oval, with 2 small central "rods" (Figure 49); *PR* = 11.5.

THORAX. Foretarsus (Figures 46 and 47) with all sensillae present except *b′ 1*; *e* and *g* spatulate, slender, pointed; *f2* half as long as *f1*; *c′* normal, reaching $\delta 6$; *a* reaching $\gamma 2$; *c* far surpassing $\gamma 3$; *t1* between $\alpha 3$ and $\alpha 3'$ (*BS* = 0.9); empodium not reaching tip of claw (*EU* = 0.9); claw small (*TR* = 5.0). Empodium of middle and hind tarsi less than one-fifth as long as claw; spine on hind tarsus distinct (Figure 50).

Chaetotaxy of metanotum: *p1′* long, behind the line *p1–2*.

ABDOMEN. Chaetotaxy of segments as in Table 4 and Figure 51.

On tergum VI *a3* missing; on tergum VII only *a4* and *a5* present; on tergum X all setae present; *p4′* missing on terga II and III, present on all other terga; *p1′* long, surpassing *p1* on terga II–VI, encroaching on hind border, but on tergum VII short, behind the line *p1–2*, and not encroaching on hind border; on tergum VIII *p1″* and *p2* not displaced anteriorly.

Squama genitalis of female (Figure 52) with caput processus like a duck's head, but curved against median line; corpus processus with a big incision; stylus long, pointed; filum processus long.

Type data. Holotype female and **allotype** male (same slide), NN, Parkes' Farm, 88-Valley, Nelson, 15 November 1970, N. A. Martin (NZAC).

Material examined. Type specimens, plus 7 non-type examples from the type locality (2 males, 5 females; NZAC, ZMKD).

— / NN.

Collected in June and November.

Habitat not recorded.

Remarks. *E. zelandicum* is closely related to *E. australicum* Womersley (see Tuxen 1964, pp. 122–123; 1967, pp. 5–6). It differs in size, *australicum* being larger (foretarsus without claw 90 μm), the structure of the pseudoculus (the two rods are not present in *australicum*), the shape of sensillae *e* and *g* (broader in *australicum*), *BS* (1.0 in *australicum*), the length of *c* and *c′* (shorter in *australicum*), and above all the position of *p1′* on tergum VII (on a line with *p1* and *p2* in *australicum*; see Figure 53).

Table 4 Abdominal chaetotaxy of *Eosentomon zelandicum*

	I	II,III	IV,V	VI	VII	VIII	IX,X	XI	Telson
Tergum	4/10	10/14	10/16	8/16	4/16	6/9	8	8	9
Sternum	4/4	6/4	6/10	6/10	6/10	7	4	8	12

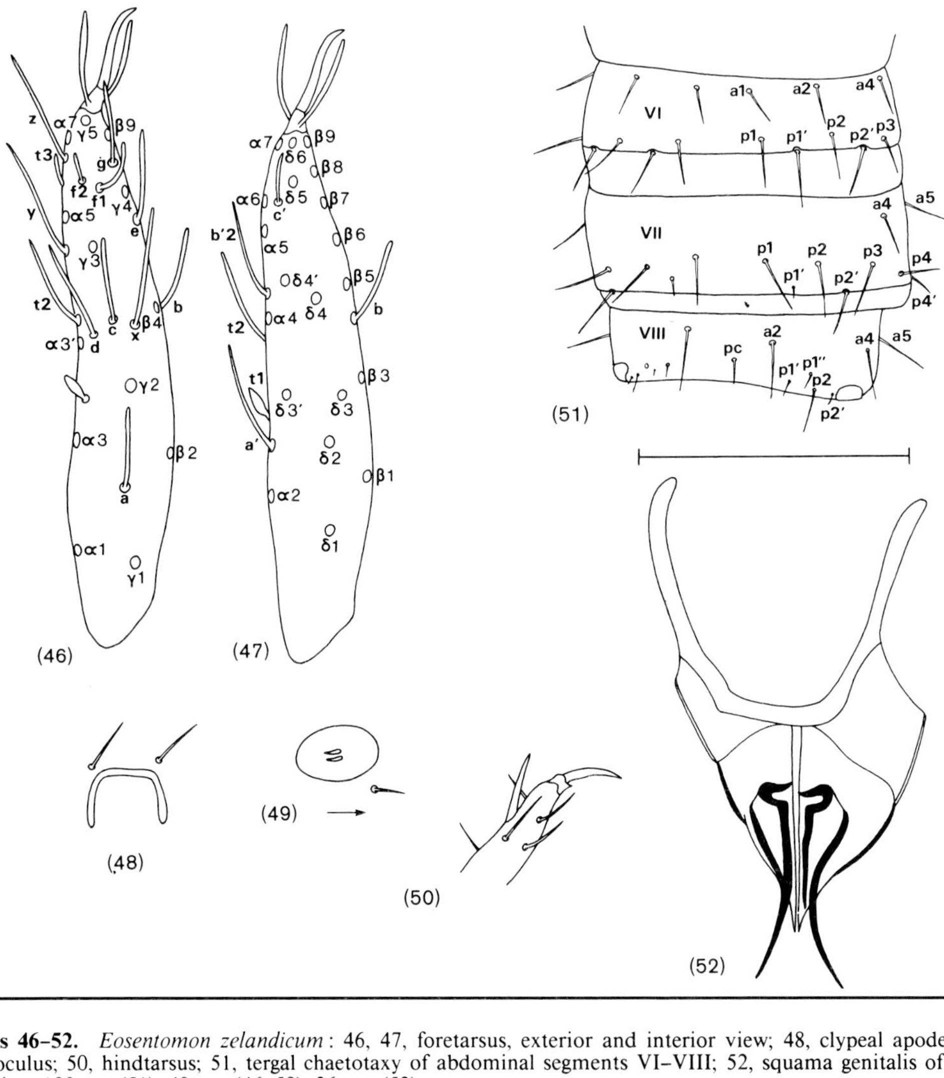

Figures 46–52. *Eosentomon zelandicum*: 46, 47, foretarsus, exterior and interior view; 48, clypeal apodeme; 49, pseudoculus; 50, hindtarsus; 51, tergal chaetotaxy of abdominal segments VI–VIII; 52, squama genitalis of female. Scale line: 100 μm (51); 40 μm (46–50); 26 μm (52).

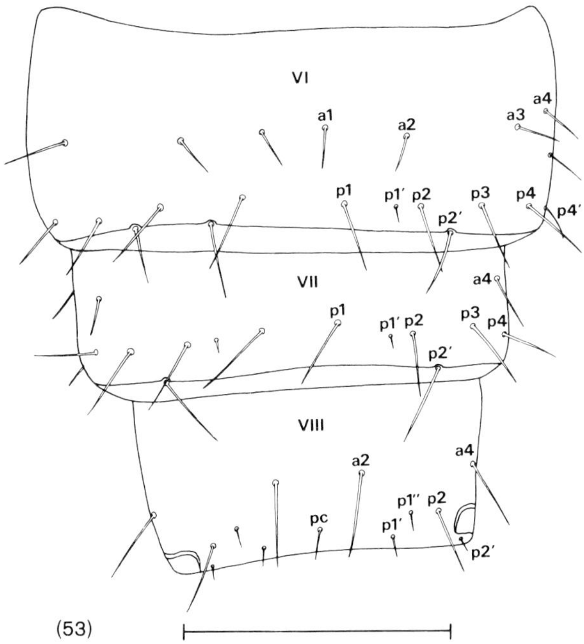

(53)

Figure 53. *Eosentomon australicum*, tergal chaetotaxy of abdominal segments VI–VIII. Scale line 100 µm.

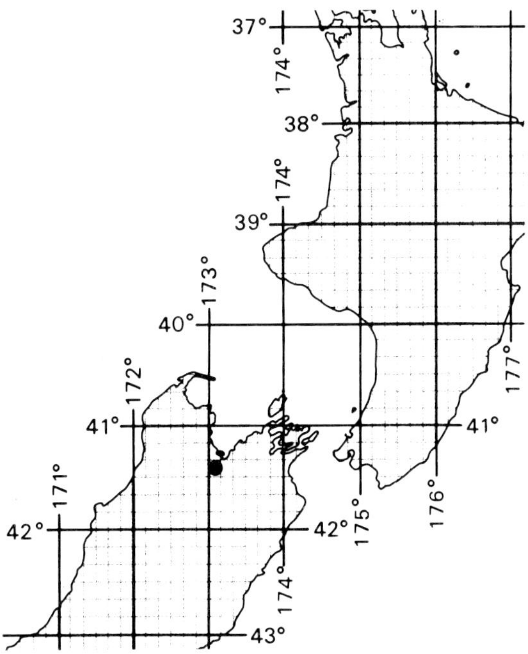

It should be noted, however, that whereas *p1′* in *zelandicum* is placed behind the line *p1–2* it does not encroach on the hind border as it does in *E. macronyx*. As to the variability of the number of anterior setae on tergum V in *australicum*, see Tuxen (1967, p. 5); it is 10 in all four New Zealand specimens. It may be that more material from the localities of both species (*australicum* is known from Australia and the U.S.A.) will show them to be identical.

This species is known only from New Zealand.

The name (adjectival) refers to the country of origin.

Eosentomon maximum new species

Figures 54–60

Length of body 1540 µm. Length of foretarsus without claw 105 µm.

HEAD. Mouthparts of the common eosentomid type. Mandibles ending in 3 small teeth. Labral setae present. Clypeal apodeme with a fairly slender anterior connection (Figure 56). Pseudoculus almost circular, with a small central globule (Figure 57); *PR* = 11–12.

THORAX. Foretarsus (Figures 54 and 55) with all sensillae present except *b′ 1*; *e* and *g* spatulate, very slender and pointed; *f2* less than half as long as *f1*; *c′* slender, behind *α6* and reaching *δ6*; *a* not reaching *γ2*; *c* hardly reaching *γ3*; *t1* between *α3* and *α3′*, but closer to the latter (*BS* = 1.1); empodium not reaching tip of claw (*EU* = 0.8); claw small (*TR* = 5.0–5.5). Empodium of middle and hind tarsi one-fifth as long as claw; spine on hind tarsus distinct (Figure 58).

Chaetotaxy of metanotum: *p1′* long, behind the line *p1–2*.

ABDOMEN. Chaetotaxy of segments as in Table 5 and Figure 59.
On terga V and VI *a3* missing; on tergum VII only *a4* and *a5* present; on tergum X all setae present; *p4′* missing on terga II and III, present on all other terga; *p1′* long, surpassing *p1* on terga II–VI, encroaching on hind border, but on tergum VII short and on a line with *p1–2*; on tegum VIII *p1″* and *p2* not displaced anteriorly.

Squama genitalis of female (Figure 60) with caput processus like a duck's head, elegantly curved against median line; corpus processus with a big incision; stylus long, pointed; filum processus long.

Type data. Holotype female and **allotype** male, NN, Dickson's Farm, The Glen, Nelson, 20 November 1971, P. G. Fenemore (NZAC).

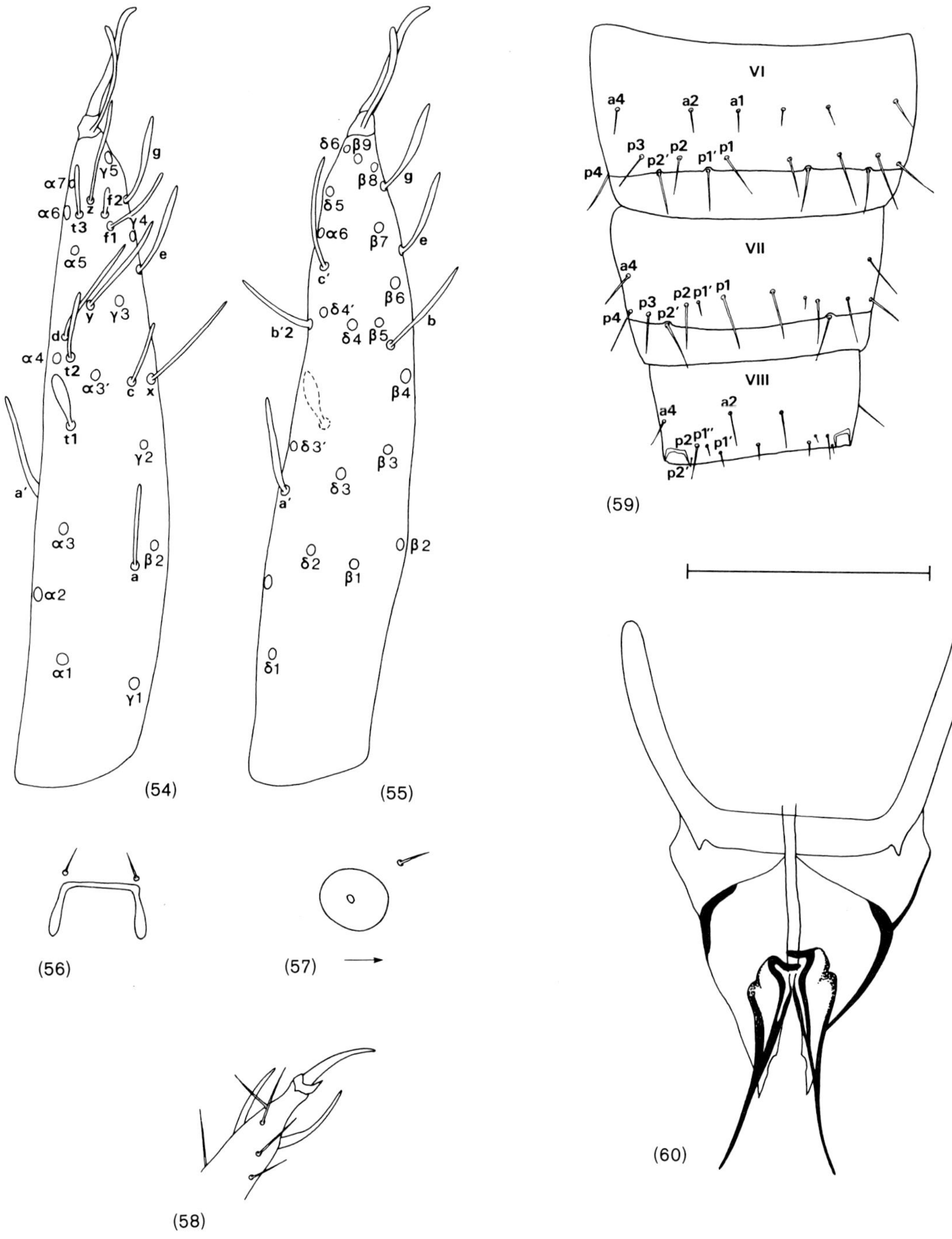

Figures 54–60. *Eosentomon maximum*: 54, 55, foretarsus, exterior and interior view; 56, clypeal apodeme; 57, pseudoculus; 58, hindtarsus; 59, tergal chaetotaxy of abdominal segments VI–VIII; 60, squama genitalis of female. Scale line: 160 µm (59); 40 µm (54–58); 26 µm (60).

Table 5 Abdominal chaetotaxy of *Eosentomon maximum*

	I	II,III	IV	V,VI	VII	VIII	IX,X	XI	Telson
Tergum	4/10	10/14	10/16	8/16	4/16	6/9	8	8	9
Sternum	4/4	6/4	6/10	6/10	6/10	7	4	8	12

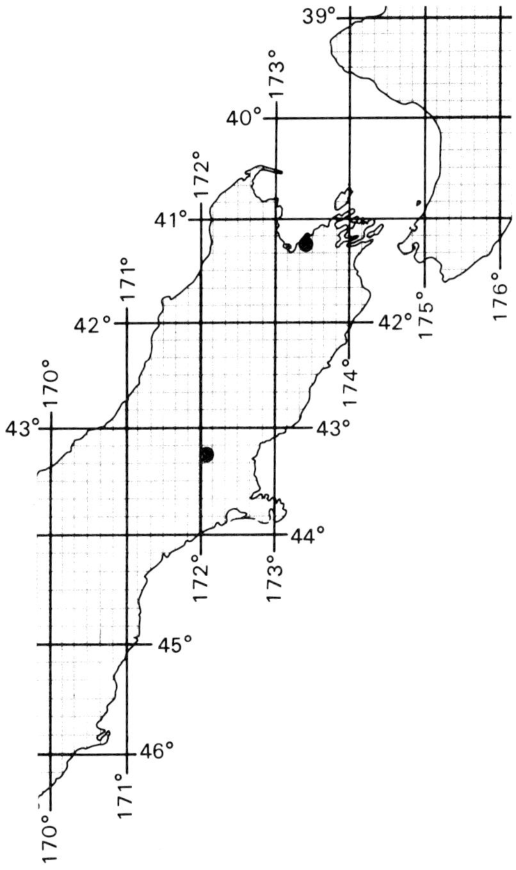

Material examined. Type specimens, plus 2 non-type females from Lees Valley beside the Ashley River (LCNZ).

— / NN, NC.

Collected in November.

Habitat unknown for the type specimens, but "mat plants on shingle beside the river" for the other two.

Remarks. This species is known only from New Zealand.

The name (adjectival) refers to its large size relative to other members of genus *Eosentomon*.

Eosentomon gracile new species

Figures 61–69

Length of body (strongly extended) 980 μm. Length of foretarsus without claw 56 μm.

HEAD. Mouthparts of the common eosentomid type. Mandibles ending in 3 small teeth. Labral setae present. Clypeal apodeme with a slender anterior connection (Figure 63). Pseudoculus oval, with a median rod or slit (Figure 64); $PR = 9.0$.

THORAX. Foretarsus (Figures 61 and 62) with all sensillae present except *b' 1*; *e* and *g* broadly spatulate, with a fairly short stem; *f2* very short; sensilla *c'* reaching *δ6*, on a line with *α6*; *a* almost reaching *γ2*; *c* reaching *γ3*; *a'* fairly slender; *t1* fairly close to *α3* ($BS = 0.9$). Empodium not reaching tip of claw ($EU = 0.9$); claw small ($TR = 5.7$). Empodium of middle and hind tarsi missing; spine on hind leg distinct (Figures 65 and 66).

Chaetotaxy of metanotum: *p1'* as long as *p1*, behind the line *p1–2*.

ABDOMEN. Chaetotaxy of segments as in Table 6 and Figure 67.

On tergum VII only *a2*, *a4*, and *a5* present; on tergum X all setae present; *p4'* missing on terga II and III, present on all other terga; *p1'* long, surpassing *p1* on terga I–VI, but short and encroaching on hind border on tergum VII; on tergum VIII *p1"* and *p2* not displaced anteriorly.

Squama genitalis of female (Figures 68 and 69) difficult to understand. In Figure 69 it is fully extended outside the body; in another specimen only a sketchy drawing of it inside the body could be made (Figure 68). Caput processus apparently elegantly curved against median line; corpus with a big incision; stylus fairly short and blunt.

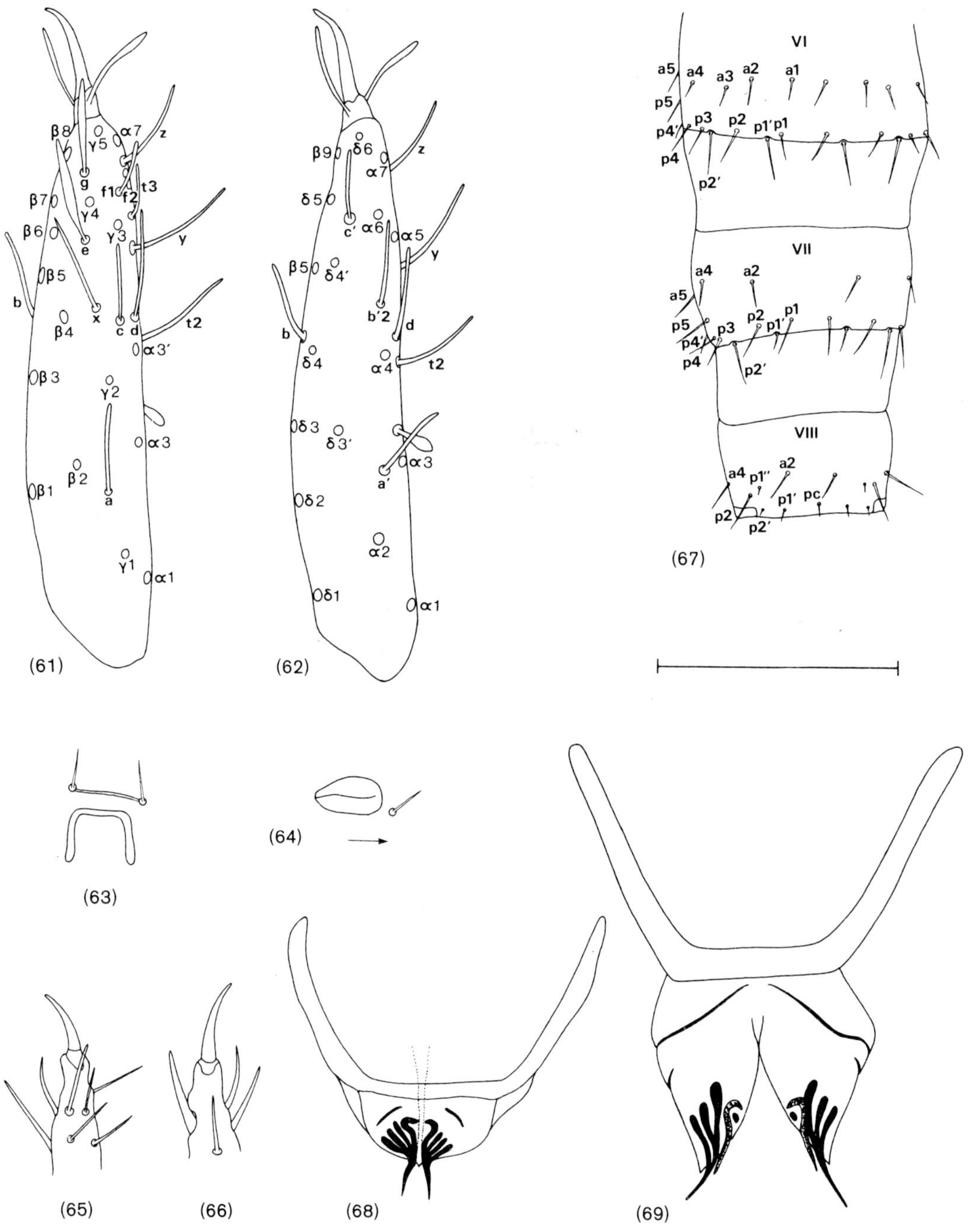

Figures 61–69. *Eosentomon gracile*: 61, 62, foretarsus, exterior and interior view; 63, clypeal apodeme; 64, pseudoculus; 65, middle tarsus; 66, hindtarsus; 67, tergal chaetotaxy of abdominal segments VI–VIII; 68, 69, squama genitalis of two different female specimens. Scale line: 100 μm (67); 26 μm (61–66, 68, 69).

Table 6 Abdominal chaetotaxy of *Eosentomon gracile*

	I	II,III	IV–VI	VII	VIII	IX,X	XI	Telson
Tergum	4/10	10/14	10/16	6/16	6/9	8	8	9
Sternum	4/4	6/4	6/10	6/10	7	6	8	12

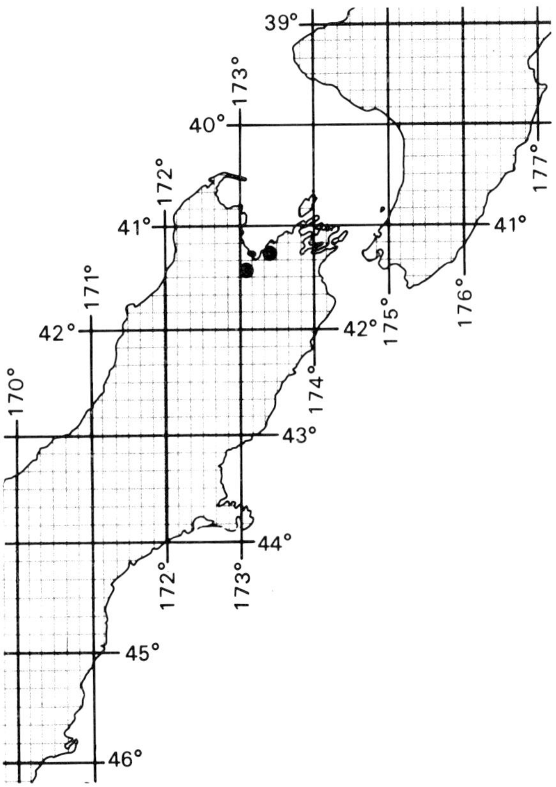

Type data. **Holotype** female and **allotype** male, NN, Parkes' Farm, 88-Valley, Nelson, 24 October 1972, N. A. Martin (NZAC).

Material examined. Type specimens, plus 15 non-type examples (9 females, 3 males, 3 maturi juniores; NZAC, ZMKD).

— / NN.

Collected in June, October, and November (pre-adults October, November).

Habitat not recorded.

Remarks. This species is known only from New Zealand.

The name (adjectival) refers to the slender body.

Suborder ACERENTOMOIDEA

Tracheal system absent. Abdominal legs more or less reduced. Claw of middle and hind legs boat-shaped. Lacinia of maxilla pointed. Pseudoculus with a rim and other structural differentiations. Abdominal tergum with a serrate line or striate band. Female squama genitalis without a processus sternalis but with acrostyli. Sternum XII with 6 or 8 setae.

Comprising two families, Acerentomidae and Protentomidae.

Family PROTENTOMIDAE

One pair (1st), two (1st and 2nd), or three (1st, 2nd, and 3rd) pairs of abdominal legs with a terminal vesicle and never more than 4 setae. Abdominal tergum VIII with a single serrate line (cf. a striate band in Acerentomidae). Female squama genitalis with bud-like or pointed acrostyli. Sternum XII with 8 setae.

Genus *Proturentomon* Silvestri

Silvestri, 1909: 10. Type species *Acerentomon minimum* Berlese, 1908, by original designation.

First and 2nd abdominal legs each with a terminal vesicle and 4 setae, 3rd without a vesicle and with 2 setae. Sensilla *t1* on foretarsus rarely missing. Metanotum with only 1 pair of anterior setae (*a2*). Lid of tergum VIII with distinct teeth. Female squama genitalis with acrostyli ending in a small, rounded knob.

***Proturentomon minimum* (Berlese)**

Figures 70–86

Berlese, 1908: 17 (*Acerentomon*). Tuxen, 1964: 182–183 (*Proturentomon*).

Length of body (extended) 900 μm. Length of foretarsus without claw 46 μm.

HEAD. Mouthparts of the protentomid type. Mandibles sturdier than those of Acerentomidae, but with a pointed apex. Maxillae with 2 slender sensillae and a terminal tuft (Figure 72). Labium with a tuft and 4 other setae (Figure 73). Canal of maxillary gland with "calyx" as a bladder and with stiffened contours on the canal (see Tuxen & Yin 1982 and Figures 74 and 75). Pseudoculus with a long "lever"; *PR* = 12 (excluding the lever); all 4 setae of pseudocular line placed between pseudoculi (Figures 76 and 77).

THORAX. Foretarsus (Figures 70 and 71) of the protentomid type, somewhat clumsy; all sensillae present except *t1*, all fairly broad, tapering towards apex, with a distinct rib; seta *s* less than half as long as claw, curved, blunt; claw with a small median tooth; *EU* = 0.15; *TR* = 3.0. Claw of middle and hind tarsi boat-shaped, with tunica lobes (Figure 78).

Chaetotaxy of metanotum: *M*, *a2*; *p1*, *p1'*, *p2*, *p3*, *p4*, *p4'* (Figure 79).

ABDOMEN. First and 2nd abdominal legs each with a terminal vesicle and 4 setae; 3rd leg without a terminal vesicle and with 2 setae, a long subapical one and a small apical, lateral one (Figures 82–84).

Chaetotaxy of segments as in Table 7 and Figures 80 and 81.

On terga I–VI only *a1* present, on tergum VII no *a* setae at all; *p4'* present on terga II–VII; *p1'* and *p3'* only on tergum VII; *p2'* never present; on terga IX and X *p4'* so small as to be easily overlooked, or perhaps absent. Comb VIII with 5 acute teeth on upper lid, about 15 very small teeth on lower lid (Figure 81). "Striate band" present as a serrate line (Figure 81).

Squama genitalis of female (Figures 85 and 86) with 2 slightly pointed, knob-like acrostyli.

Type data. Lectotype female, Italy, Florence, Giardino Boboli, 1907, A. Berlese (Stazione di Entomologia Agraria, Firenze; see Tuxen, 1960, p. 302).

Material examined. Eleven non-type females from Parkes' Farm, 88-Valley, Nelson (NZAC, ZMKD).
— / NN.

Collected in June and November.

Habitat not recorded.

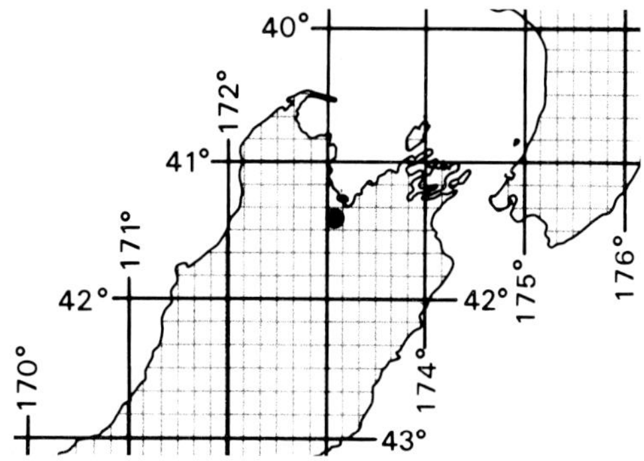

Table 7 Abdominal chaetotaxy of *Proturentomon minimum*

	I	II,III	IV–VI	VII	VIII	IX	X	XI	Telson
Tergum	2/10	2/12	2/12	0/16	6/14	12(14)	8 (10)	6	9
Sternum	4/2	4/3	4/6	2/6	4	4	4	6	8

Figures 70–86. *Proturentomon minimum* : 70, 74, foretarsus, exterior and interior view; 72, maxillary palp; 73, labial palp; 74, 75, canal of maxillary gland, ventral and lateral view; 76, pseudoculus; 77, pseudoculi, and setae of pseudocular line; 78, middle tarsus; 79, tergal chaetotaxy of thoracic segment III; 80, tergal chaetotaxy of abdominal segment V; 81, tergal chaetotaxy of abdominal segments VII and VIII; 82–84, abdominal legs I, II, and III; 85, 86, squama genitalis of female, dorsal and lateral view. Scale line: 100 μm (79–80); 64 μm (81); 40 μm (70, 71, 77, 78, 82–86); 26 μm (72–76).

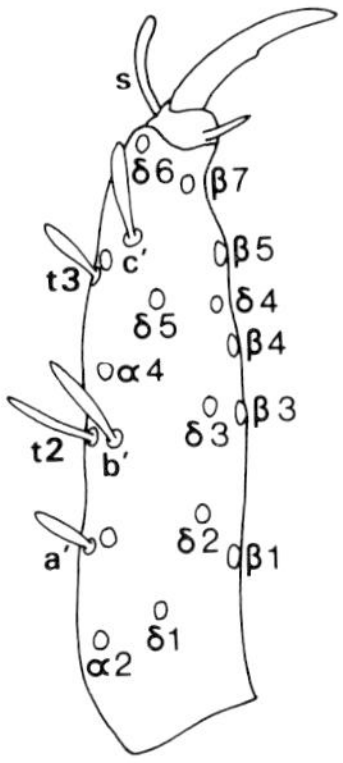

(70)

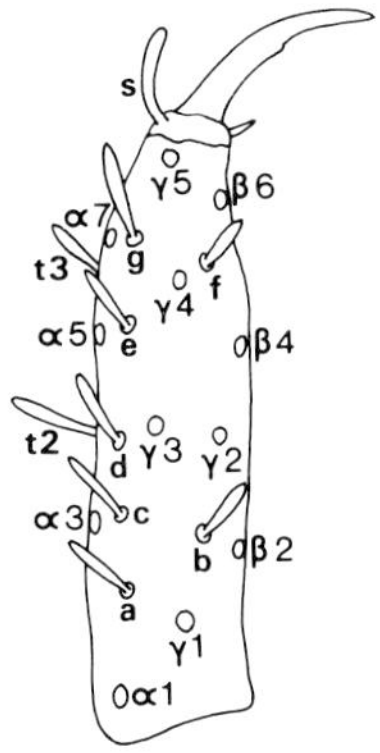

(71)

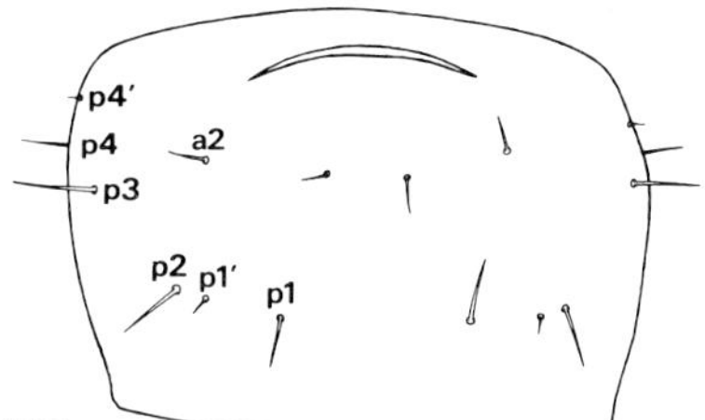

(79)

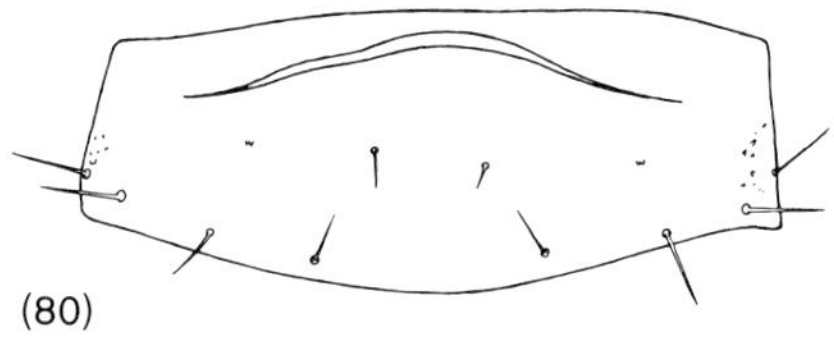

(80)

(72)

(73)

(74)

(75)

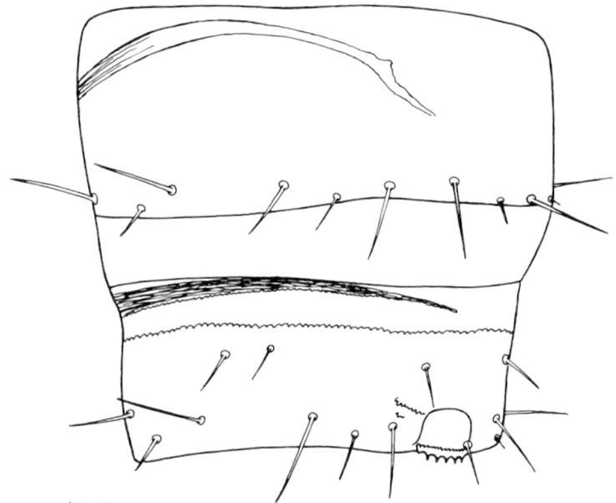

(81)

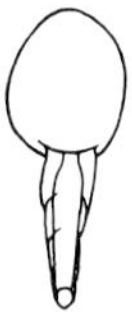

(76)

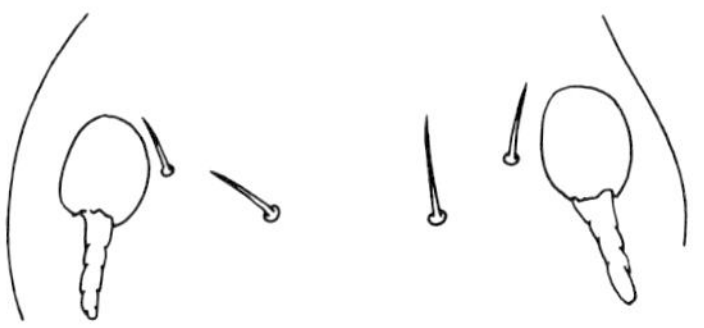

(77)

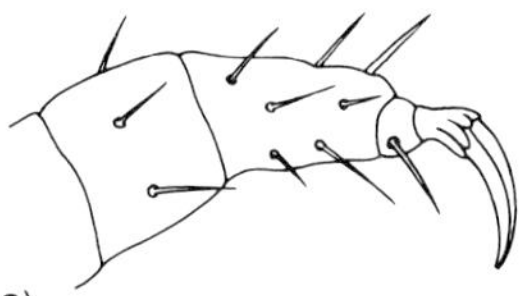

(78)

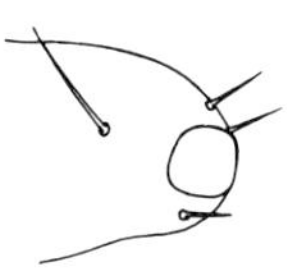

(82)

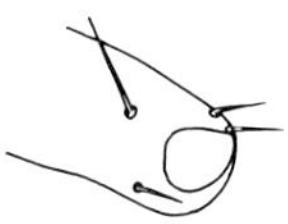

(83)

(84)

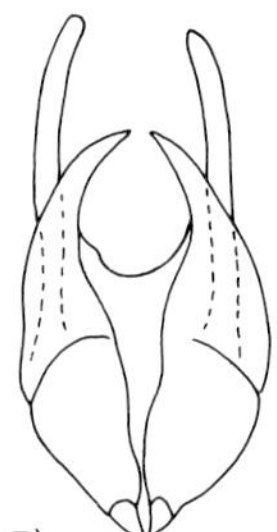

(85)

(86)

Remarks. No males of *P. minimum* are known, from anywhere in its range. It was probably introduced to New Zealand, where it is known only from the above record.

P. minimum has also been recorded from England and central and southern Europe. The genus is known from Europe, North Africa, and the U.S.A. (Iowa), but not (except for this record) from the Southern Hemisphere; *Proturentomon acracia* Vidal & Najt, recorded from Argentina, is actually a *Protentomon* (Tuxen 1978a, p. 36).

The name refers to the unusually small size of this species.

Family ACERENTOMIDAE

Only 1st abdominal leg with a terminal vesicle and 4 setae, 2nd and 3rd without a vesicle and each with 3 setae, 2, or 1. Abdominal tergum VIII with a striate band consisting of 2 lines, with or without striae in between. Female squama genitalis more or less pointed, bipartite or, together with distal part of stylus, forming a tripartite structure. Sternum XII with 6 setae.

Genus *Acerentulus* Berlese

Berlese, 1908: 122. Type species *Acerentomon confine* Berlese, 1908, by original designation.

Acerentomidae with striate band well developed, with a regular grate of stakes. Labial palp not reduced, with a terminal tuft of setae. Second and 3rd abdominal legs each with 3 setae — a long subapical seta, a shorter lateral apical, and a delicate median apical. Sensilla *t1* on foretarsus claviform, *t3* willow leaf-shaped, *b'* present; Sternum VIII with posterior setae.

Acerentulus kermadecensis Ramsay & Tuxen

Figures 87–96

Ramsay & Tuxen, 1978: 603–606.

Length of body 1250–1500 μm. Length of foretarsus without claw 105 μm.

HEAD without a rostrum. Mouthparts of the form common in *Acerentulus*. Maxillary palp with a tuft and 2 sensillae (Figure 89). Labial palp with a tuft, 3 setae, and a pointed sensilla (Figure 90). Canal of maxillary gland with a narrow calyx expanded distally into a small excrescence; proximal part as long as proximal branch of fulcrum, its ending bipartite (Figure 91). Pseudoculus longer than broad, with a lever resembling a strong handle (Figure 92); *PR* = 12.5–14.0.

THORAX. Foretarsus (Figures 87 and 88) with all sensillae present; *t1* club-shaped, the club sometimes very narrow; *t3* long, awl-shaped; *a* reaching *d*, but far from *γ*; *b* and *c* almost equal in length; *d* extremely long, longer than *e* or *g*; *a'* shaped as a narrow sword; *b'* extremely long, reaching even *α7*; *c'* exceptionally long; empodium small, pointed (*EU* = 0.15); *BS* = 0.5–0.6 (on one foreleg of holotype 0.9; in view of this aberration I have drawn the foreleg of another specimen); *TR* = 3.7–4.0.

Chaetotaxy of metanotum: *M*, *a2*, *a4*; *p1*, *p1'*, *p2*, *p2'*, *p3*, *p4*, *p4'*, *p5* (microchaeta).

ABDOMEN. Second and 3rd abdominal legs each with 3 setae — a long subapical, a smaller apicolateral, and a still smaller median apical one (Figures 93 and 94).

Chaetotaxy of segments as in Table 8.

On terga II–V *a1*, *a2*, and *a5* present; *a4* added to these on terga VI and VII; *p1'* missing on terga I–VI, present on tergum VII; *p4'* present on terga II–VII; *p3'* present on tergum VII. Abnormalities found: an additional *pc* on tergum VII; only 4 setae in hind row of sternum III. Striate band on segment VIII well developed, with distinct striae filling out the entire space; comb VIII with about 10 fairly long teeth (Figure 95).

Squama genitalis of female with pointed acrostyli (Figure 96).

Type data. Holotype female, New Zealand, Kermadec Islands, Raoul Island, Mount Moumoukai, 427 m, in litter beneath *Rhopalostylis cheesemanii*, 23 January 1967, J.C. Watt (NZAC).

Material examined. Holotype, plus 8 non-type examples (6 females, 1 maturus junior, 1 larva I; NZAC, ZMKD).

Kermadec Is / — / NN.

Collected in January, June, November, and December (pre-adult stages: larva I — November; maturus junior — December).

Taken from soil under liverworts, moss, and *Hymenophyllum*.

Figures 87–96. *Acerentulus kermadecensis*: 87, 88, foretarsus, exterior and interior view; 89, maxillary palp; 90, labial palp; 91, canal of maxillary gland; 92, pseudoculus; 93, 94, abdominal legs II and III; 95, tergal chaetotaxy of abdominal segment VIII; 96, squama genitalis of female (redrawn from Ramsay & Tuxen 1978). Scale line: 40 μm (87–95); 30 μm (96).

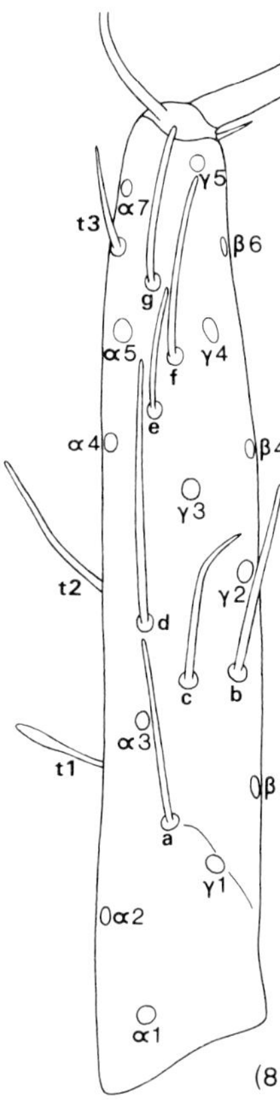

(87)

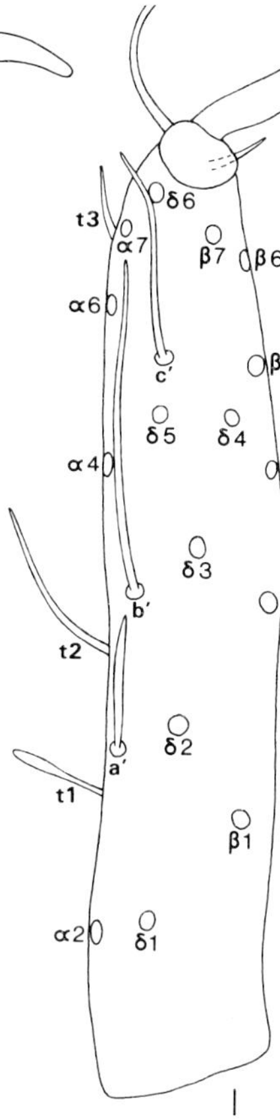

(88)

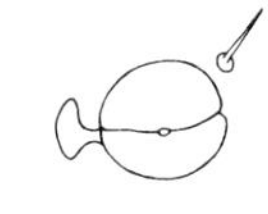

(92)

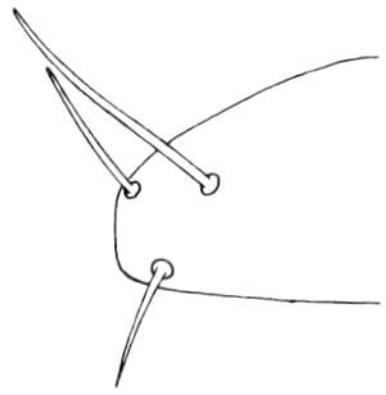

(93)

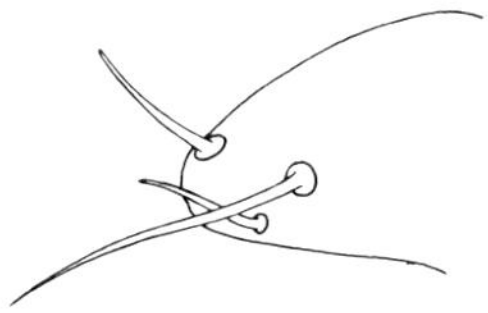

(94)

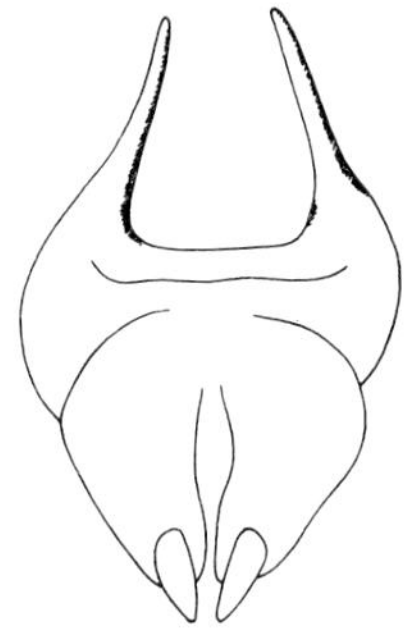

(96)

(89)

(90)

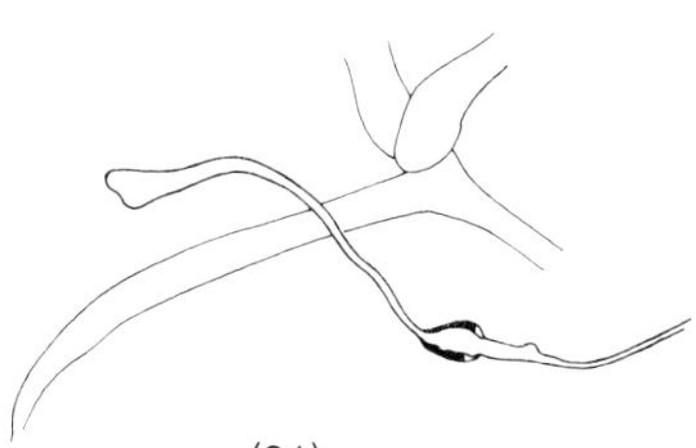

(91)

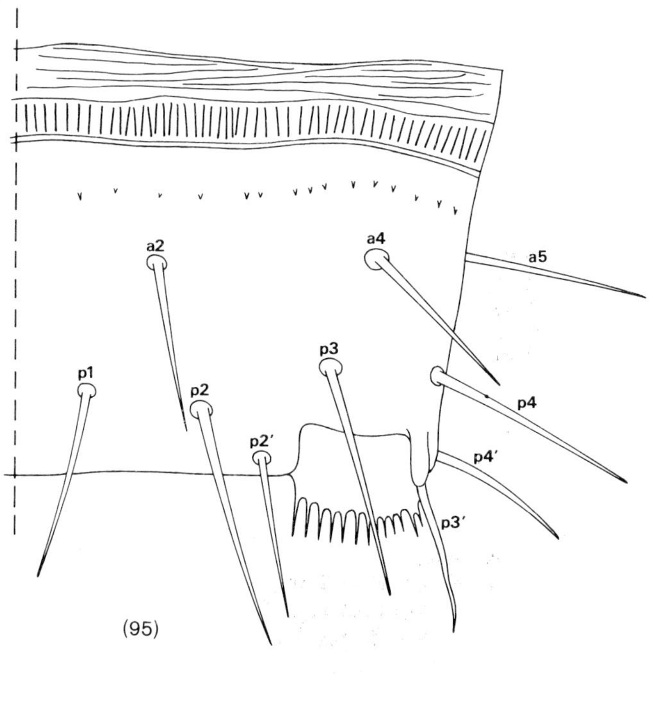

(95)

Table 8 Abdominal chaetotaxy of *Acerentulus kermadecensis*

	I	II,III	IV,V	VI	VII	VIII	IX,X	XI	Telson
Tergum	6/10	6/14	6/14	8/14	8/18	6/16	12	6	9
Sternum	3/4	3/5	3/8	3/8	3/8	4/2	4	6	6

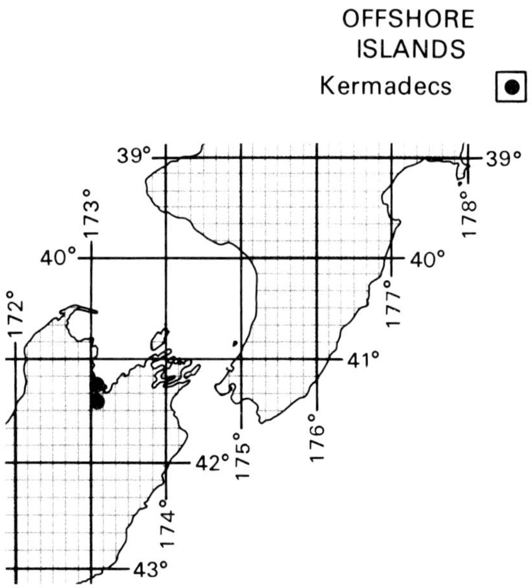

Remarks. Only females of *A. kermadecensis* are known. The holotype is an abnormal specimen, and the drawings of it in the original description came out rather badly, so the drawings given here are from another specimen. The apparent distribution within New Zealand is very curious (see *Berberentulus nelsoni*, p. 46).

This species is known only from New Zealand. The name refers to the type locality.

Genus *Australentulus* Tuxen

Tuxen, 1967: 18. Type species *Acerentulus australiensis* Womersley, 1932, by original designation.

Acerentomidae with striate band developed, with or without a hook-shaped design. Labial palp reduced, with 3 setae. Second and 3rd abdominal legs each with 3 setae, as in *Acerentulus*. Sensilla *t1* on foretarsus claviform, *t3* bud-shaped, *b′* present. Sternum VIII without posterior setae.

Australentulus sp. indet.

Only one specimen has been found of this genus, which in Australia is common and represented by many species; it is also known from the Malay Peninsula, India, and Madagascar. Unfortunately the specimen is in ecdysis from larva I to larva II, which makes identification of setae and sensillae impossible. The striate band is visible, but whether complete or not cannot be seen; possibly it has a hook-shaped design. The three setae on the second and third abdominal legs are distinctly visible, as also is the reduction of the labial palp.

The collection date are: KA, Oaro, in a large tree hole, 12 November 1969, R. M. Emberson (LCNZ).

Genus *Gracilentulus* Tuxen

Tuxen 1963: 89. Type species *Acerentulus gracilis* Berlese, 1908, by original designation.

Acerentomidae with striate band well developed, with a regular grate of striae. Labial palp reduced, with 3 setae. Second and 3rd abdominal legs each with 2 setae, a long subapical seta and a delicate median apical one. Sensilla *t1* on foretarsus claviform, *t3* a small knob, *b′* present or absent. Sternum VIII without posterior setae.

Gracilentulus gracilis (Berlese)

Figures 97–106

Berlese, 1908: 122 (*Acerentulus*). Tuxen, 1964: 296–297 (*Gracilentulus*).

Length of body (extended) 1000 µm. Length of foretarsus without claw 77 µm.

HEAD. Mouthparts of the acerentomid type. Maxillary palp with a tuft and 2 pointed sensillae (Figure 99). Labial palp with 3 setae and a fairly short sensilla (Figure 100). Canal of maxillary gland with an oval calyx, indistinct distal excrescences, and a proximal part less than half as long as proximal branch of fulcrum, broadening proximally

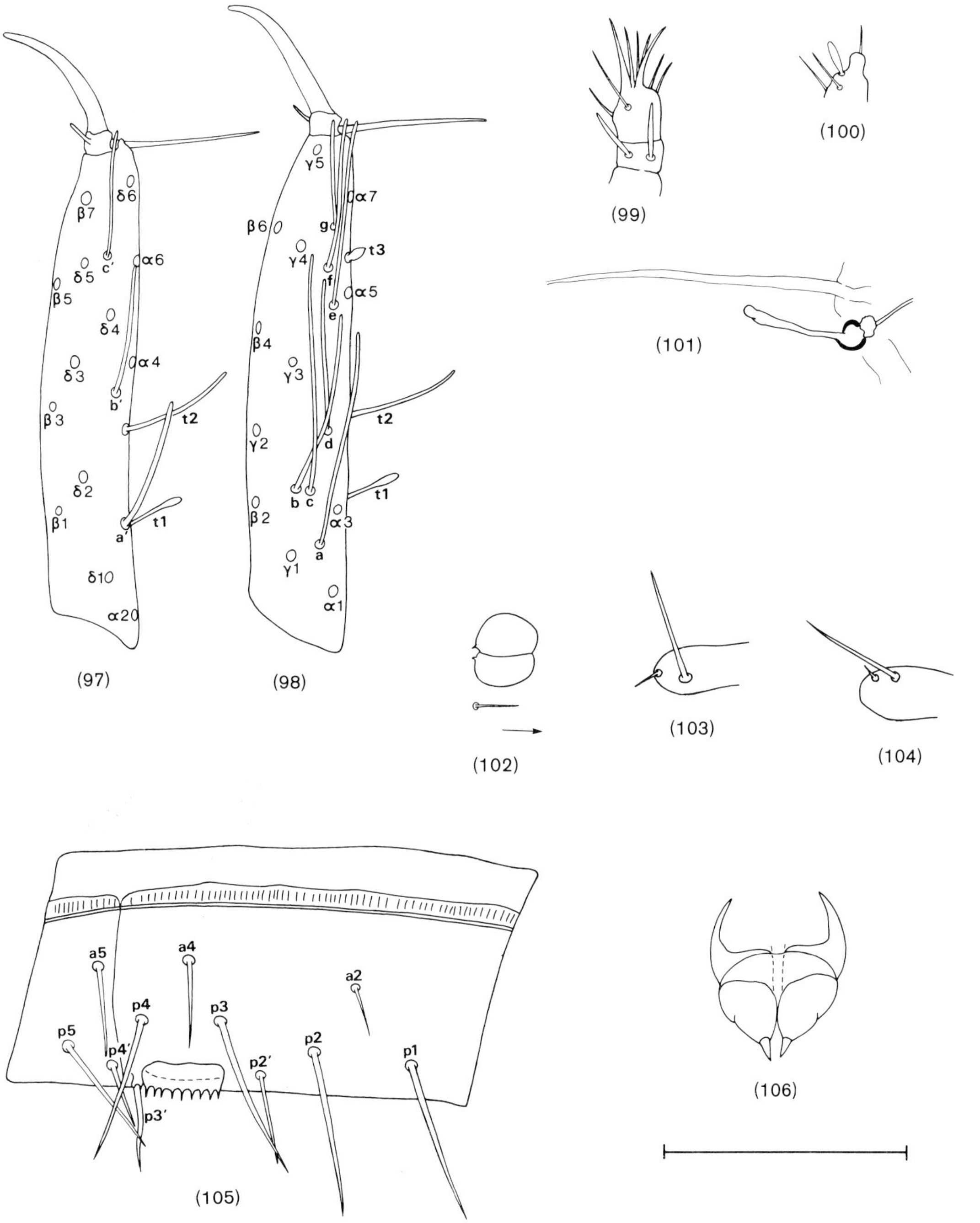

Figures 97–106. *Gracilentulus gracilis*: 97, 98, foretarsus, exterior and interior view; 99, maxillary palp; 100, labial palp; 101, canal of maxillary gland; 102, pseudoculus; 103, 104, abdominal legs II and III; 105, tergal chaetotaxy of abdominal segment VIII; 106, squama genitalis of female. Scale line: 40 μm (97, 98, 103–106); 26 μm (99–102).

Table 9 Abdominal chaetotaxy of *Gracilentulus gracilis*

	I	II,III	IV–VI	VII	VIII	IX	X	XI	Telson
Tergum	6/10	8/14	8/14	8/16	6/16	14	12	6	9
Sternum	3/4	3/5	3/8	3/8	4	4	4	6	6

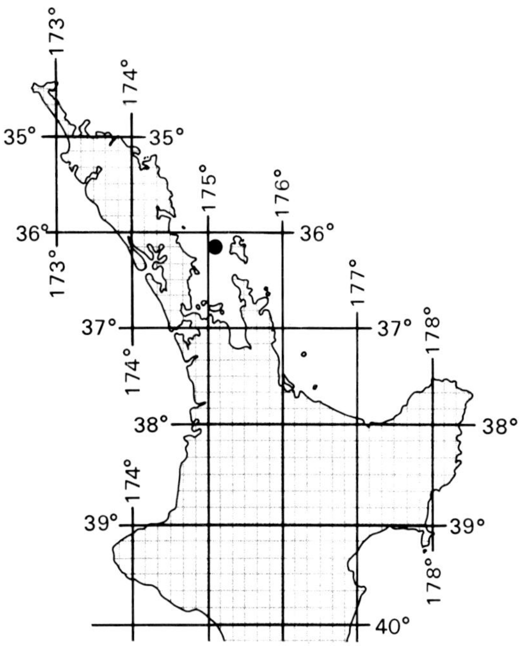

(Figure 101). Pseudoculus broader than long (Figure 102); *PR* = 14.

THORAX. Foretarsus (Figures 97 and 98) with all sensillae present and very long; *t1* claviform, the club fairly narrow; *t3* knob-shaped; *a′* narrow sword-shaped, reaching *b′*; *a* and *b* reaching *e*; *c* reaching *γ4*; *d* reaching *f*; *e*, *f*, *g*, and *c′* surpassing base of claw; *BS* = 0.40; *EU* = 0.20; *TR* = 3.2.

Chaetotaxy of metanotum: *a2*, *a4*, *M*; *p1*, *p1′*, *p2*, *p2′*, *p3*, *p4*, *p5*.

ABDOMEN. First abdominal leg with a terminal vesicle and 4 setae; 2nd and 3rd legs without a terminal vesicle, with a long subapical seta and a median apical seta less than half as long as the subapical one (Figures 103 and 104).

Chaetotaxy of segments as in Table 9.

Terga II–VII with *a1*, *a2*, *a4*, and *a5* present; *p1′* only on tergum VII; *p4′* on terga II–VII. (NOTE. On some tergites the chaetotaxy is variable — the anterior row of tergum VII has a central seta (in all thus 7 *a*) in 5 of the 7 New Zealand specimens, and the posterior row of tergum VIII has a central seta (in all thus 15 *p*) in 2 of them. Such variability is also known from other material of the species. The chaetotaxy of tergum IX is here given as 14; it was given as 12 in Tuxen (1964), and this error was repeated by Nosek (1973). All the European specimens in ZMKD, all adult specimens from Madeira, and the only known specimen from Australia (see below) have 14 setae on tergum IX; these include a smaller and a larger pleural seta, although the smaller one is missing in the maturus junior stage.) Striate band well developed, with a distinct grate. Comb VIII (the opening of the big abdominal glands) straight, with 9 or 10 teeth (Figure 105).

Squama genitalis of female (Figure 106) with pointed acrostyli and short basal arms.

Type data. **Lectotype** female, Italy, Toiana, Pisa (Stazione di Entomologia Agraria, Firenze; see Tuxen 1964, p. 297).

Material examined. Seven non-type females from Little Barrier Island (NZAC, ZMKD).

CL (Little Barrier I.) / —.

Collected in October.

Habitat details not recorded — "Pohutukawa Flat foreshore".

Remarks. Only females of *G. gracilis* are known from the Southern Hemisphere, and males seem altogether to be rare.

This species has also been recorded from Europe except the northern part (Nosek 1973), Madeira (Tuxen 1982), Morocco (Nosek 1973), South Africa (Condé 1955), and Australia (the material cited by Tuxen (1967) was an erroneous identification of *Berberentulus aestuarii* Tuxen, but recently I have seen one specimen, from South Australia).

The name presumably refers to the slender body shape.

Genus *Amphientulus* Tuxen

Tuxen, 1981: 137. Type species *Berberentulus validus* Tuxen, 1967, by original designation.

Acerentomidae with striate band reduced, without any trace of striae, but often with a hook-shaped design. Labial palp reduced, with 3 setae. Second and 3rd abdominal legs each with 2 setae, a long subapical seta and a strong lateral apical one. Sensilla *t1* on foretarsus claviform, *t3* jar-shaped, *b'* present, *b* most often shorter than *c*. Sternum VIII with or without posterior setae.

Amphientulus zelandicus new species

Figures 107–118

Length of body (extended) 950 µm; length of foretarsus without claw 92 µm.

HEAD. Maxillary palp with a tuft and 2 tapering sensillae (Figure 109). Labial palp reduced, with 4 setae and a sausage-shaped sensilla (Figure 110). Canal of maxillary gland with a heart-shaped calyx and distal extensions (excrescences); proximal part almost as long as proximal branch of fulcrum, proximally with 3 globulose dilatations (Figure 111; holotype and several other specimens) or simply bifid (Figure 112; most specimens). Pseudoculus almost circular, a little broader than long (Figure 113); $PR = 15$.

THORAX. Foretarsus (Figures 107 and 108) with all sensillae present, most of them extremely long; *t1* claviform; *t3* short jar-shaped; *a* surpassing *t2*; *b*, *c*, and *d* reaching or surpassing *f*; *f* and *g* surpassing base of claw; *a'* surpassing *b'*; *c'* reaching base of claw; $BS = 0.43$; $EU < 0.10$; $TR = 3.0$.

Chaetotaxy of metanotum: *a2*, *a4*, *M*; *p1*, *p1'*, *p2*, *p2'*, *p3*, *p4*, *p5*.

ABDOMEN. First abdominal leg with a terminal vesicle and 4 setae; 2nd and 3rd legs without a terminal vesicle, but with a long subapical seta and a fairly long apicolateral seta (Figures 114 and 115).

Chaetotaxy of segments as in Table 10.

Terga II–IV with *a1*, *a2*, and *a5* present; *a4* added to these on terga V and VI; *a2*, *a4*, and *a5* on tergum VII (and exceptionally a central, asymmetrically placed seta); *p1'* present on terga I–VII; *p4'* on terga II–VII.

Striate band reduced, but tergum with a distinct hook-shaped design. Comb VIII oblique, with fairly large teeth. Number of glands on segments VII and VIII: 9 + 9 dorsomedially, 6 + 6 laterally, 4 + 4 ventrally (Figure 116).

Squama genitalis of female (Figures 117 and 118) with long, pointed acrostyli each bearing a small, square, median tooth.

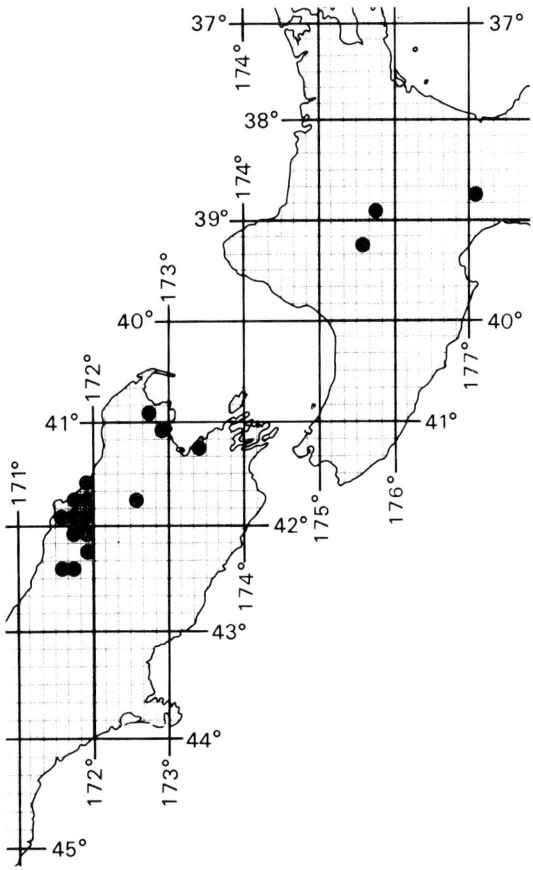

Type data. Holotype female, TO, Tokaanu, Te Ponanga Bush, 27 January 1976, B. M. May (NZAC).

Material examined. Holotype, plus some hundreds of non-type examples representing all life stages (LCNZ, NZAC, ZMKD).

TO, GB / NN, MB, BR.

Collected in January, March, April, and June–November (pre-adult stages: maturi juniores — April, July, September–November; larvae II — September, November (1 in ecdysis); larvae I and prelarvae — November).

Highest recorded elevation 1070 m (Mount Ruapehu, TO).

Taken from litter and soil under moss, Podocarpaceae, *Nothofagus*, *Metrosideros*, *Weinmannia*, and many other plants, with no apparent preferences; and in a nest of the ant *Huberia striata*.

Remarks. *A. zelandicus* differs from most of the species hitherto described in its genus in having sensilla *b* as long as *c*, a condition shared only by *gnangarae* Tuxen from Australia and *ruseki* Nosek

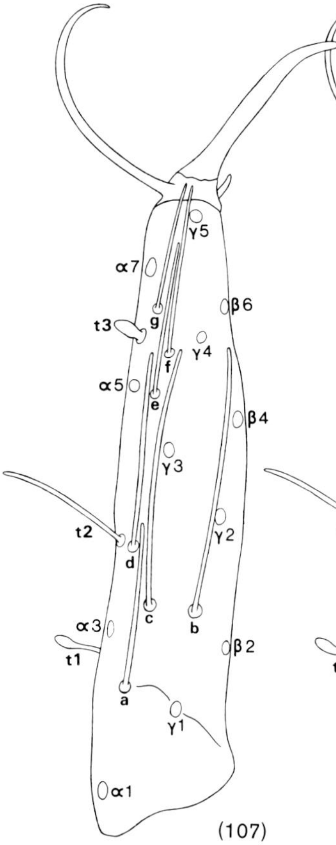

(107)

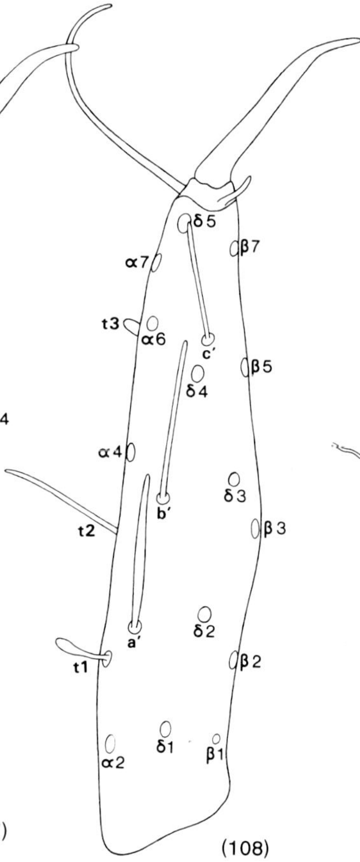

(108)

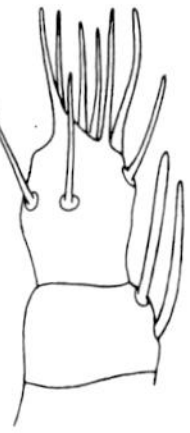

(109)

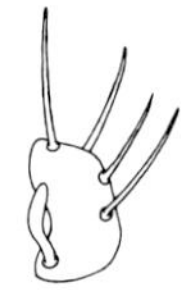

(110)

(111)

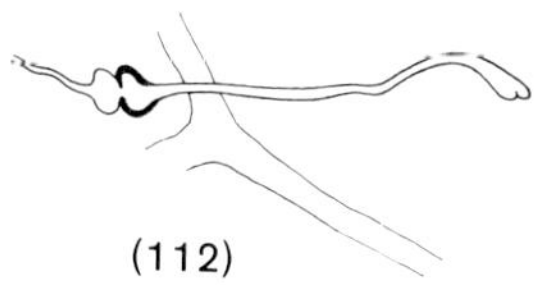

(112)

(113)

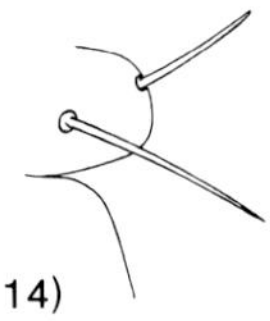

(114)

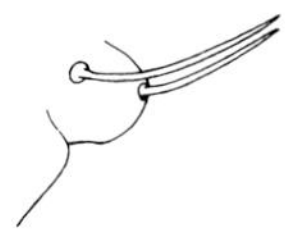

(115)

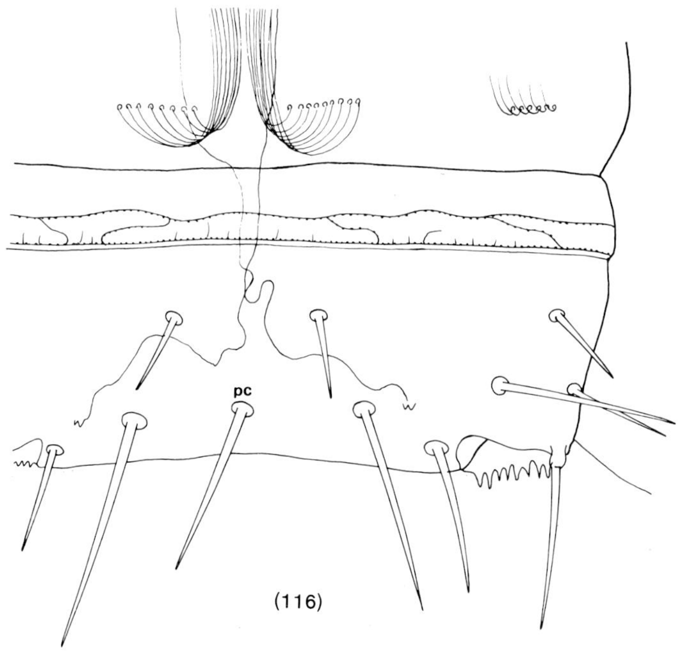

(116)

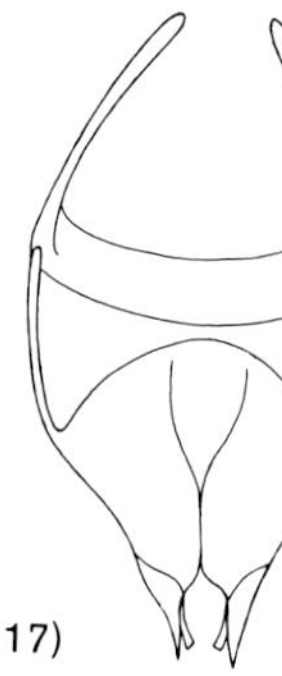

(117)

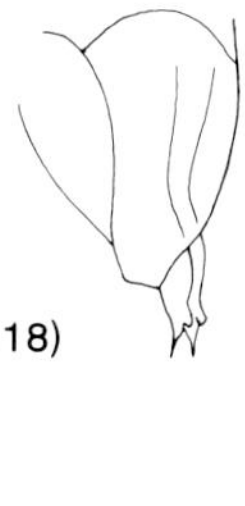

(118)

Figures 107–118. *Amphientulus zelandicus*: 107, 108, foretarsus, exterior and interior view; 109, maxillary palp; 110, labial palp; 111, 112, canal of maxillary gland (111, holotype); 113, pseudoculus; 114, 115, abdominal legs II and III; 116, tergal chaetotaxy of abdominal segment VIII; 117, 118, squama genitalis of female, dorsal and lateral view. Scale line: 40 μm (107, 108, 111–118); 26 μm (109, 110).

Table 10 Abdominal chaetotaxy of *Amphientulus zelandicus*

	I	II,III	IV	V,VI	VII	VII	IX	X	XI	Telson
Tergum	$\frac{6}{12}$	$\frac{6}{16}$	$\frac{6}{16}$	$\frac{8}{16}$	$\frac{6}{16}$	$\frac{6}{15}$	14	12	6	9
Sternum	$\frac{3}{4}$	$\frac{3}{5}$	$\frac{3}{6}$	$\frac{3}{8}$	$\frac{3}{8}$	4	4	4	6	6

from Madagascar. It differs from both these in the presence of *p1′* on all abdominal terga and in the length of the maxillary gland canal.

This species is known only from New Zealand.

The genus is represented by several species in Australia; it is recorded dubiously from Madagascar, and with still more doubt from Korea.

The name refers to the country of origin.

Tasmanentulus new genus

"aff. *Gracilentulus*" Tuxen, 1981: 131 and 138. Type species *Gracilentulus tasmanicus* Tuxen, 1967.

(The name is derived from that of the type species, in accordance with the practice for many other acerentomid genera; gender masculine.)

Acerentomidae with striate band reduced (the striae may be seen more or less distinctly, but never as a distinct grate — see below). Labial palp reduced, with 4 setae and a sensilla. Second and 3rd abdominal legs each with 2 setae, the apical one lateral and more than half as long as the subapical one. Sensilla *t1* on foretarsus claviform, *t3* more or less parallel-sided, *b′* present or absent. Sternum VIII with 2 posterior setae (occasionally only 1 in maturus junior). Stylus in female genitalia fairly clumsy; acrostylus small, pointed.

Remarks. In 1967 I described two species from Australia, *tasmanicus* and *similis*, under the genus *Gracilentulus*, but with the notion that they differed from other species of this genus in having two posterior setae on sternum VIII and a "striate band reduced to some degree though the striae are still easily seen". In my key to acerentomid genera (Tuxen 1981) I overlooked this reservation, and though I indicated the striae in the bands with dotted lines (figures 26 and 27, p. 133) I regarded the band as "complete" and arranged the two species accordingly as "aff. *Gracilentulus* Tx." (p. 131, and no. 21 in the key). The ample material from New Zealand shows that, even though the striae are visible, a separation must be made between a distinct cuticular grate and the weaker "striae" which in reality are probably internal glandular canals (see Tuxen 1981, p. 127). The difference is clearly apparent in the scanning electron micrographs (Figures 10 and 12) of an *Amphientulus* and a *Gracilentulus*. Regarded in this way the "genus" aff. *Gracilentulus* in my 1981 key must be removed to no. 29 in the key, resembling *Amphientulus* but distinguished by the two posterior setae on sternum VIII. A hook-shaped design is present in the striate band, as in probably all species of *Amphientulus*.

The Australian species *Tasmanentulus tasmanicus* (Tuxen) **new combination** and *T. similis* (Tuxen) **new combination** may be distinguished from the New Zealand representative of the genus, described below, by means of the following key.

01 Sensilla *a′* on foretarsus broad; pseudoculus broad (*PR* = 18) ... ***intermedius***
— Sensilla *a′* slender; pseudoculus oval (*PR* = 14–16) ... 2

02(01) Canal of maxillary gland long; sensilla *d* between *c* and *t2*; comb VIII with 14 or 15 teeth ... ***tasmanicus***
— Canal of maxillary gland short; sensilla *d* level with *t2*; comb VIII with 9 or 10 teeth ... ***similis***

Tasmanentulus intermedius new species

Figures 119–129

Length of body (extended) 1100 µm; length of foretarsus without claw 105 µm.

HEAD. Mouthparts of the common acerentomoid type. Maxillary palp with a tuft and 2 long, pointed sensillae on penultimate segment (Figure 121). Labial palp with 4 setae and a sausage-like sensilla (Figure 122). Canal of maxillary gland varying in length and in shape of calyx (Figure 123, holotype; Figures 124 and 125 at higher magnification, showing 2 extremes). Pseudoculus broader than long (Figure 126); *PR* = 18.

THORAX. Foretarsus (Figures 119 and 120) with all sensillae present, and almost all very long; *t1* claviform; *t3* short, parallel-sided; *b* only half as long as *c*; *a'* broader than other sensillae, surpassing base of *b'*; *BS* = 0.5; *EU* = 0.10; *TR* = 3.0.

Chaetotaxy of metanotum: *a2, a4, M*; *p1, p1', p2, p2', p3, p4, p5*.

ABDOMEN. First abdominal leg with a terminal vesicle and 4 setae; 2nd and 3rd legs without a terminal vesicle, but with a long subapical seta and a fairly long apicolateral seta (Figure 127).

Chaetotaxy of segments as in Table 11.

Terga II–V with *a1, a2,* and *a5* present; *a4* added to these on tergum VI; *ac, a2, a4,* and *a5* on tergum VII (exceptionally *a1, a2, a4,* and *a5* or even *a2, a4,* and *a5*); *p1'* present on terga I–VII; *p4'* on terga II–VII.

Striate band reduced, but tergite with a distinct hook-shaped design. Comb VIII with 7–9 fairly large teeth (Figure 128).

Squama genitalis of female (Figure 129) with basal arms short; acrostyli broad, clumsy, ending in a point and furnished with a tooth as in *Amphientulus*; a double papilla between acrostyli.

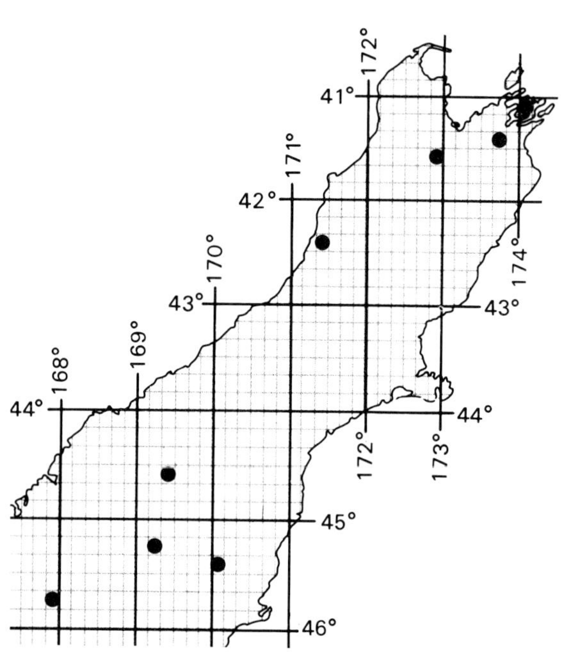

Type data. Holotype female: CO, Old Man Range, 1432 m, in *Chionochola pallens*, 24 February 1974, J.C. Watt (NZAC).

Material examined. Holotype, plus 50 non-type examples (27 females, 14 males, 7 maturi juniores, 2 larvae II; NZAC, ZMKD).

— / NN, SD, MB, BR, CO, FD, SI.

Collected in January–April, July, September, and November (pre-adult stages: maturi juniores — September; larvae II — March).

Highest recorded elevation 1585 m (Gordons Knob, NN).

Table 11 Abdominal chaetotaxy of *Tasmanentulus intermedius*

	I	II,III	IV,V	VI	VII	VIII	IX	X	XI	Telson
Tergum	6/12	6/16	6/16	8/16	7/16	6/16	14	12	6	9
Sternum	3/4	3/5	3/8	3/8	3/8	4/2	4	4	6	6

Figures 119–129. *Tasmanentulus intermedius*: 119, 120, foretarsus, exterior and interior view; 121, maxillary palp; 122, labial palp; 123–125, canal of maxillary gland (123, holotype); 126, pseudoculus; 127, abdominal leg III; 128, tergal chaetotaxy of abdominal segment VIII; 129, squama genitalis of female. Scale line: 40 µm (119, 120, 123, 126–129); 26 µm (121, 122, 124, 125).

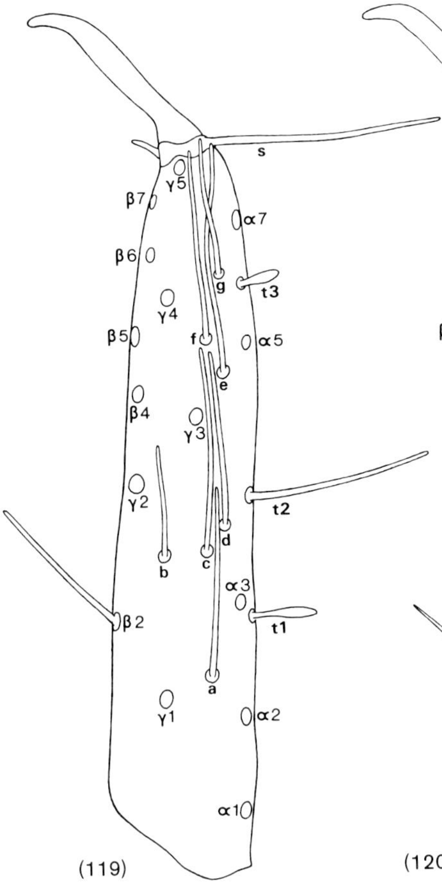

(119)

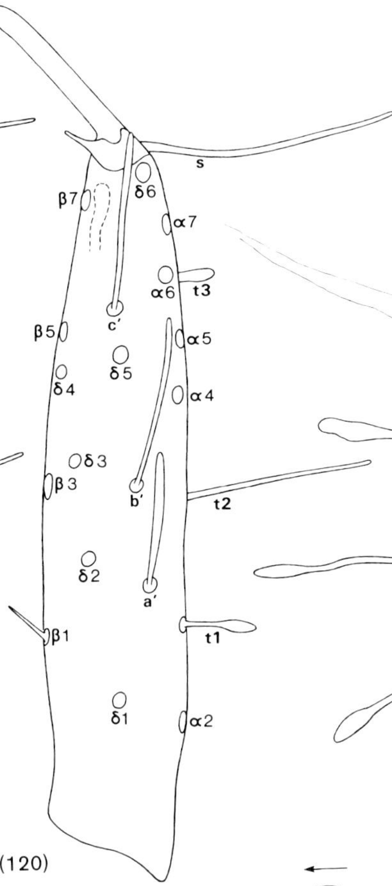

(120)

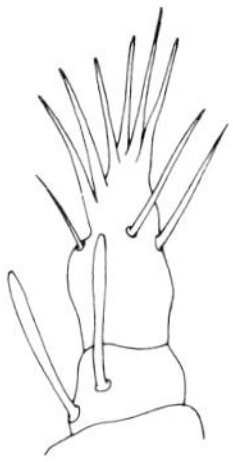
(121)

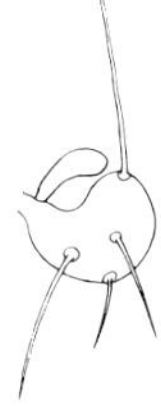
(122)

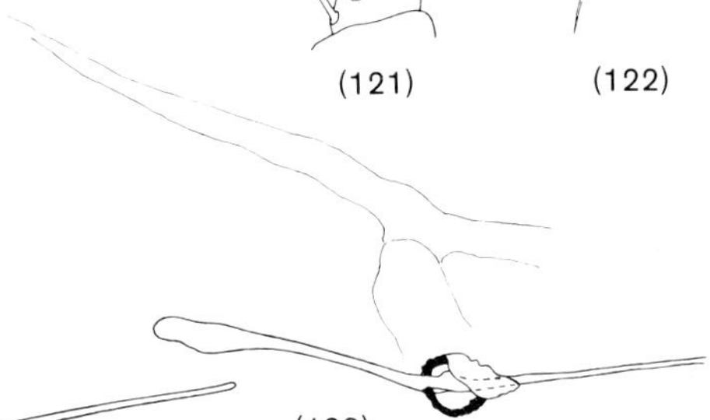
(123)

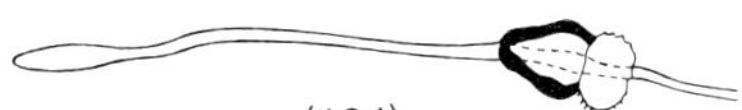
(124)

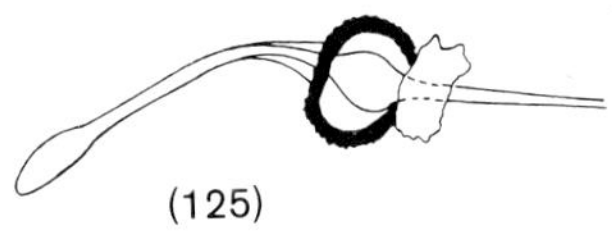
(125)

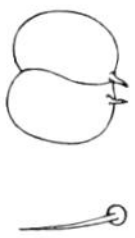
(126)

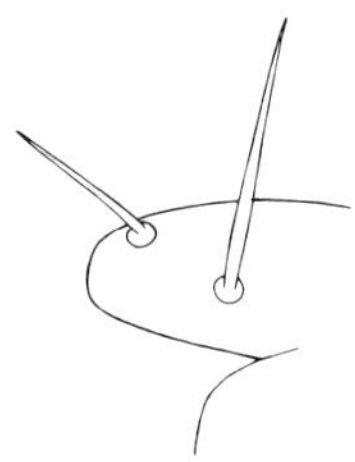
(127)

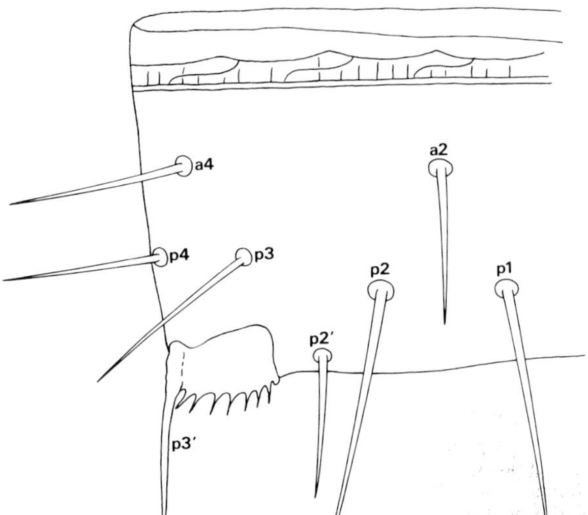

(128)

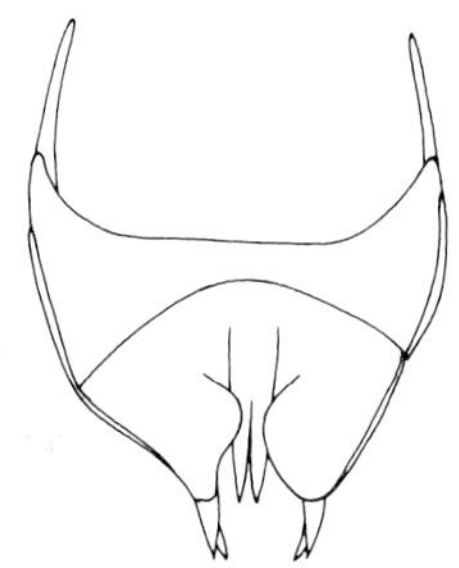
(129)

Taken from soil and litter under *Nothofagus*, *Metrosideros*, *Weinmannia*, etc. at lower elevations, but above about 1000 m from swards with *Celmisia*, *Marsippospermum*, and *Schoenus*. At Mawhera (BR) taken together with *Amphientulus zelandicus*, but alone farther south and at greater heights, and hence more cold-tolerant than this closely related species.

Remarks. The trivial name refers to the systematic position of *intermedius* between *tasmanicus* and *similis*.

Yinentulus new genus

Type species *Yinentulus paedocephalus* new species.

(The name is given in honour of Professor Yin Wen-ying, of Shanghai, for her remarkable ability to find unique Protura; gender masculine.)

Acerentomidae with striate band reduced. Labial palp reduced, with 3 setae and a sensilla. Second and 3rd abdominal legs each with 2 setae, the apical one lateral and shorter than the subapical one. Sensilla *t1* on foretarsus baculiform; sensilla *b* distal to *c–d*, extremely long. (NOTE: Whether the curious shape of the head (see below) is a generic character remains to be seen.)

Yinentulus paedocephalus new species

Figures 130–146

Length of body (extended) 840 μm; length of foretarsus without claw 60 μm.

HEAD. Mouthparts of the common acerentomid type. Maxillary palp with a tuft and 2 pointed sensillae (Figure 132). Labial palp reduced, with 3 setae and a sausage-shaped sensilla (Figure 133). Canal of maxillary gland with a heart-shaped calyx, distal extensions, and a proximal part (Figure 134), less than half as long as proximal branch of fulcrum. Pseudoculus circular (Figure 135); *PR* = 13. (NOTE: The most remarkable feature is the shape of the head, which in profile has a prominent "front" like the head of a young child, not the common slender shape of other acerentomids. Figure 136 shows the head of the holotype in profile; Figures 137 and 138 show the head of a maturus junior in dorsal and ventral view.)

THORAX. Foretarsus (Figures 130 and 131) with all sensillae present; *t1* baculiform; *t3* expanded distally, almost baculiform; *b* distal to the line *c–d*, extremely long; *a*, *e*, and *f* very long; *a'* sword-shaped; *c'* long; *EU* = 1.1; *BS* = 0.5; *TR* = 2.3. Claw of 2nd and 3rd legs boat-shaped, with a long empodium and tunica lobes (Figure 139).

Chaetotaxy of metanotum: *a2*, *a4*, *M*; *p1*, *p1'*, *p2*, *p2'* close to *p3*, *p4*, *p5*.

ABDOMEN. First abdominal leg (Figure 140) with a terminal vesicle and 4 setae; 2nd and 3rd legs with a long subapical seta and a shorter apicolateral one (Figure 141); exception — 3 setae on 2nd leg of holotype (Figure 142).

Chaetotaxy of segments as in Table 12.

Terga II–VII with *a1*, *a2*, and *a5* present; *a2*, *a4*, and *a5* on tergum VII; *p1'* absent on terga I–VI, present on tergum VII; *p4'* present on terga II–VII.

Striate band on tergum VIII reduced; comb VIII with many very small teeth (Figure 145).

Table 12 Abdominal chaetotaxy of *Yinentulus paedocephalus*

	I	II,III	IV–VI	VII	VIII	IX,X	XI	Telson
Tergum	6/10	6/14	6/14	6/16	6/15	12	4	9
Sternum	3/4	3/5	3/8	3/8	4	4	6	6

Figures 130–146. *Yinentulus paedocephalus*: 130, 131, foretarsus, exterior and interior view; 132, maxillary palp; 133, labial palp; 134, canal of maxillary gland; 135, pseudoculus; 136, head, holotype, lateral view; 137, 138, head, maturus junior, dorsal and ventral view; 139, hindtarsus; 140–144, abdominal legs, holotype (140, leg I; 141, 142, leg II; 143, 144, leg III); 145, tergal chaetotaxy of abdominal segment VIII; 146, squama genitalis of female, lateral view. Scale line: 40 μm (130, 131, 136–145); 26 μm (132–135, 146).

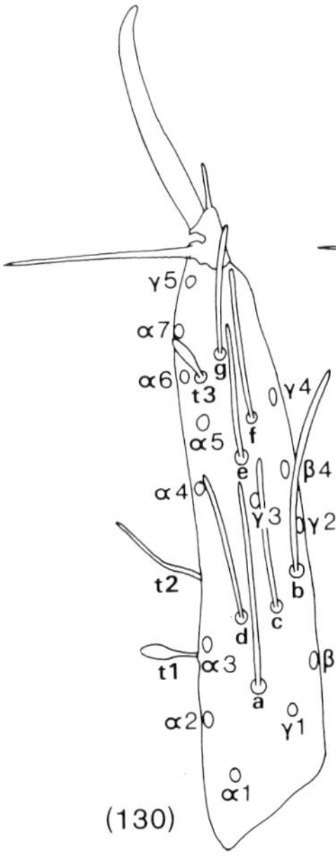

(130)

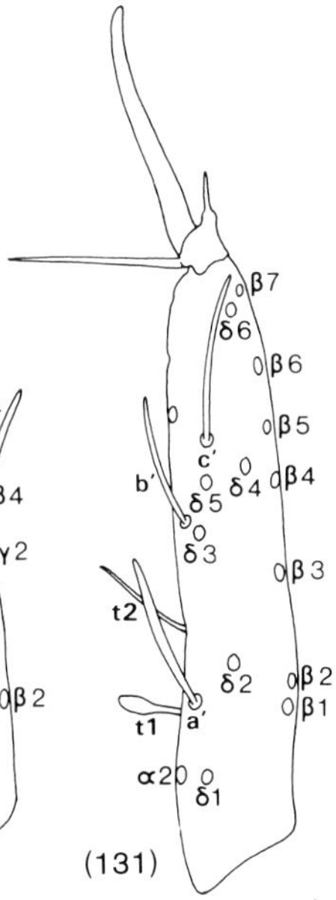

(131)

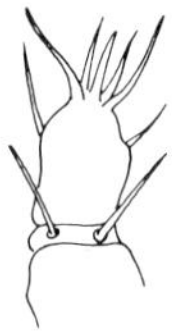

(132)

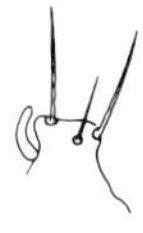

(133)

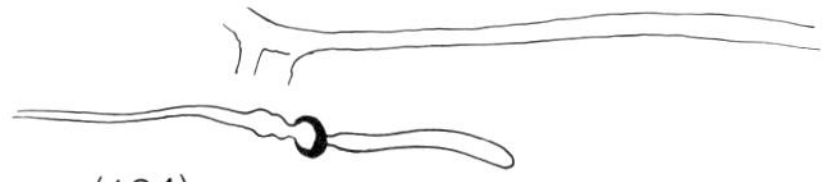

(134)

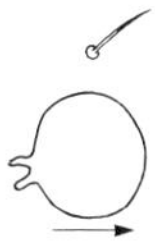

(135)

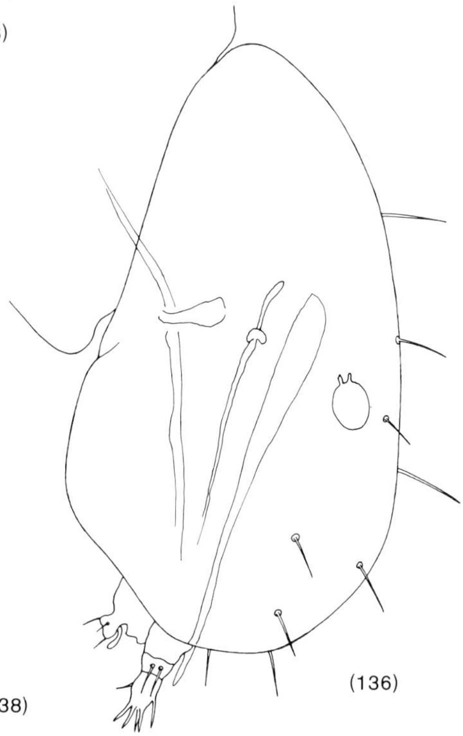

(136)

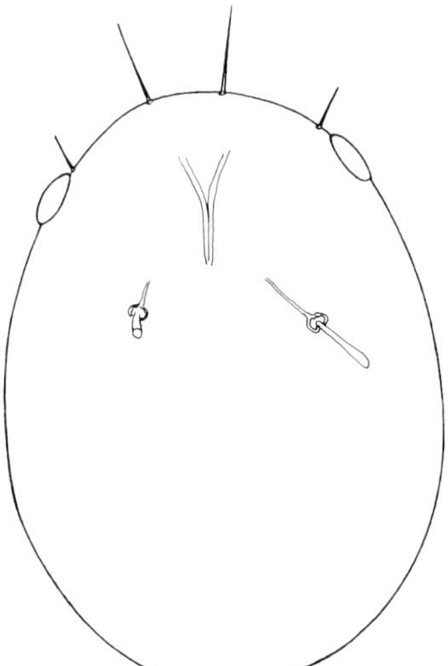

(137)

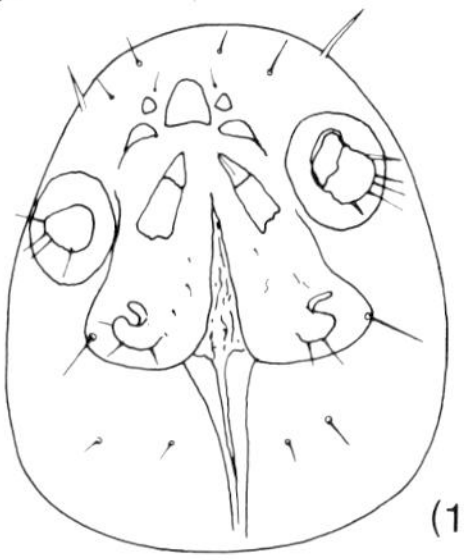

(138)

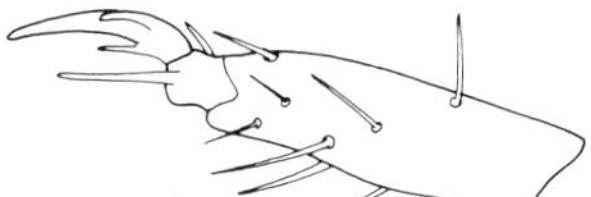

(139)

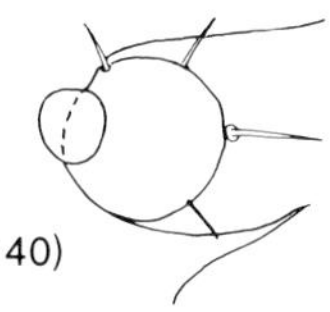

(140)

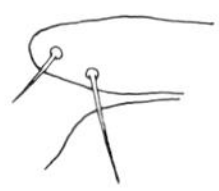

(141)

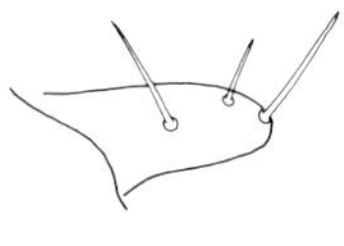

(142)

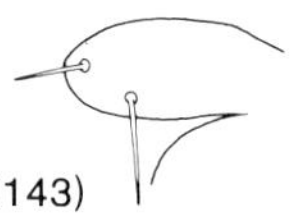

(143)

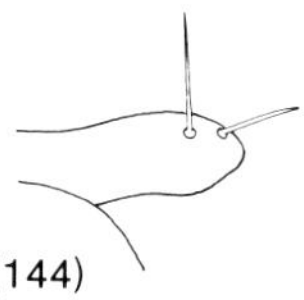

(144)

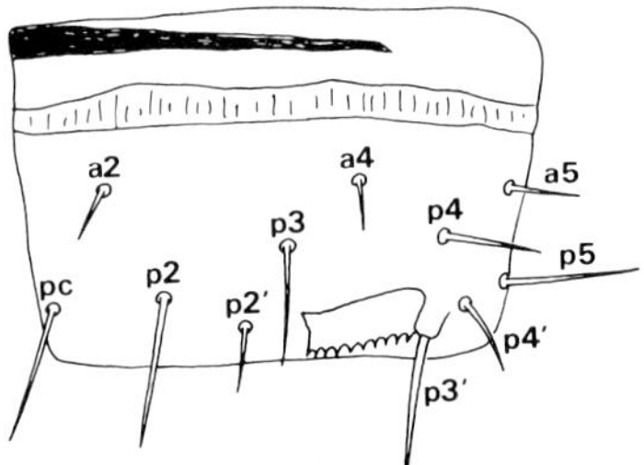

(145)

(146)

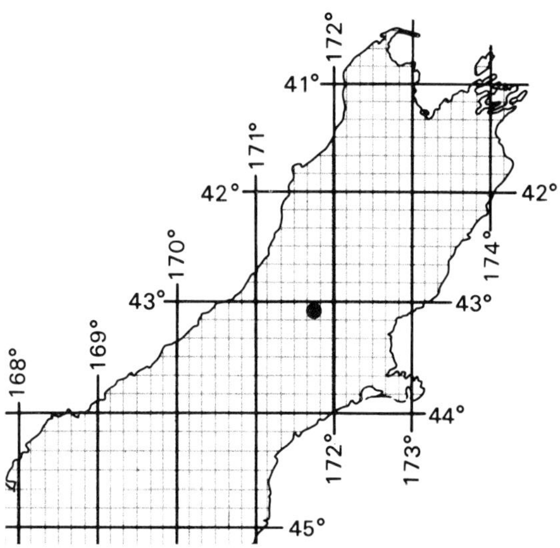

Squama genitalis of female (Figure 146) with pointed acrostyli. (NOTE: On all slides this structure is very difficult to understand, since all specimens are oriented laterally in fairly thick mounts. This is also the reason for several other doubtful elements in the above description.)

Type data. Holotype female: MC, Craigieburn Forest Park, north side of Mount Wall, *ca* 1200 m, litter of *Nothofagus solandri cliffortiodes*, 21 May 1971, R.M. Emberson (NZAC).

Material examined. Holotype, plus 3 non-type examples collected with it (2 females, 1 maturus junior; LCNZ, ZMKD).

Remarks. The name refers to the shape of the head, and is derived from the Greek *pais, paidós* (child) and *kephalé* (head).

Genus *Kenyentulus* Tuxen

Tuxen, 1981: 135. Type species *Acerentulus kenyanus* Condé, 1948, by original designation.

Acerentomidae with striate band reduced, the striae only very faintly visible in phase contrast. Labial palp reduced, with 3 setae. Maxillary gland with 2 or 3 dilatations proximal to calyx, cf. none in all other acerentomid genera from New Zealand. Second and 3rd abdominal legs each with 2 setae, a long subapical one and a very delicate median apical one. Sensilla *t1* on foretarsus baculiform, *t3* knob-shaped, *b'* present; sternum VIII without posterior setae.

Kenyentulus kenyanus (Condé)

Figures 147–155

Condé, 1948: 749–751 (*Acerentulus*). Tuxen, 1964: 300–301 (*Gracilentulus*); –1978b: 256–257 and key (*Gracilentulus*); –1981: 135 and 139 (*Kenyentulus*).

Length of body 770 µm; length of foretarsus without claw 52 µm.

HEAD. Mouthparts of the common acerentomid type. Maxillary palp with a tuft and 2 thin, pointed sensillae (Figure 149). Labial palp reduced, with 3 setae and a sausage-shaped sensilla (Figure 150). Canal of maxillary gland with an oval calyx and 2 dilatations besides the proximal one (Figure 151). Pseudoculus circular; 4 setae in pseudocular line, all between the pseudoculi (Figure 152); *PR* = 15.

THORAX. Foretarsus (Figures 147 and 148) with all sensillae present; *a, f,* and *c'* broader than the others; *a* reaching *γ3*; *b* extremely short, hardly surpassing *γ2*; *a'* sword-like; *t1* baculiform, the knob slender but distinct; *t3* knob-like; *BS* = 0.45; *EU* = 0.15; *TR* = 3.3.

Chaetotaxy of metanotum: *a2, a4, M*; *p1, p1', p2, p2', p3, p4.*

ABDOMEN. First abdominal leg with a terminal vesicle and 4 setae; 2nd and 3rd legs (Figures 153 and 154) with a long subapical seta and a very small median apical one.

Chaetotaxy of segments as in Table 13.

Terga II–VI with *a1, a2,* and *a5* present; *a2, a4,* and *a5* on tergum VII; *p1'* present on terga I–VII; *p4'* on terga II–VII; *p3'* on tergum VII only; tergum IX with only 12 setae, not 14 as stated by Tuxen (1964, p. 300).

Striate band reduced, with some striae weakly visible. Comb VIII with 8 small teeth (Figure 155).

Squama genitalis of female not seen; in examples from elsewhere it resembles that of *Gracilentulus gracilis.*

Type data. Lectotype female: Kenya, Kisumu, 18 June 1947 (collection of Prof. B. Condé, Nancy, France).

Material examined. Two non-type examples — 1 female, 1 maturus junior, AK, Lynfield, around roots of *Metrosideros excelsa*, 6 May 1980, G. Kuschel (NZAC).

Remarks. *K. kenyanus* has also been recorded from Kenya, the Seychelles, southern India (Kerala), Brazil, Puerto Rico (Nosek 1978, with doubt, because *a'* and *b'* on the interior side of the foretarsus are shorter than stated by Tuxen (1964); however, it is hardly justified to attach a specific

Table 13 Abdominal chaetotaxy of *Kenyentulus kenyanus*

	I	II,III	IV–VI	VII	VIII	IX,X	XI	Telson	
Tergum	6/12	6/16	6/16	6/18	6/15	12	6	9	
Sternum	3/4	3/5	3/8	3/8	4	4	4	4	6

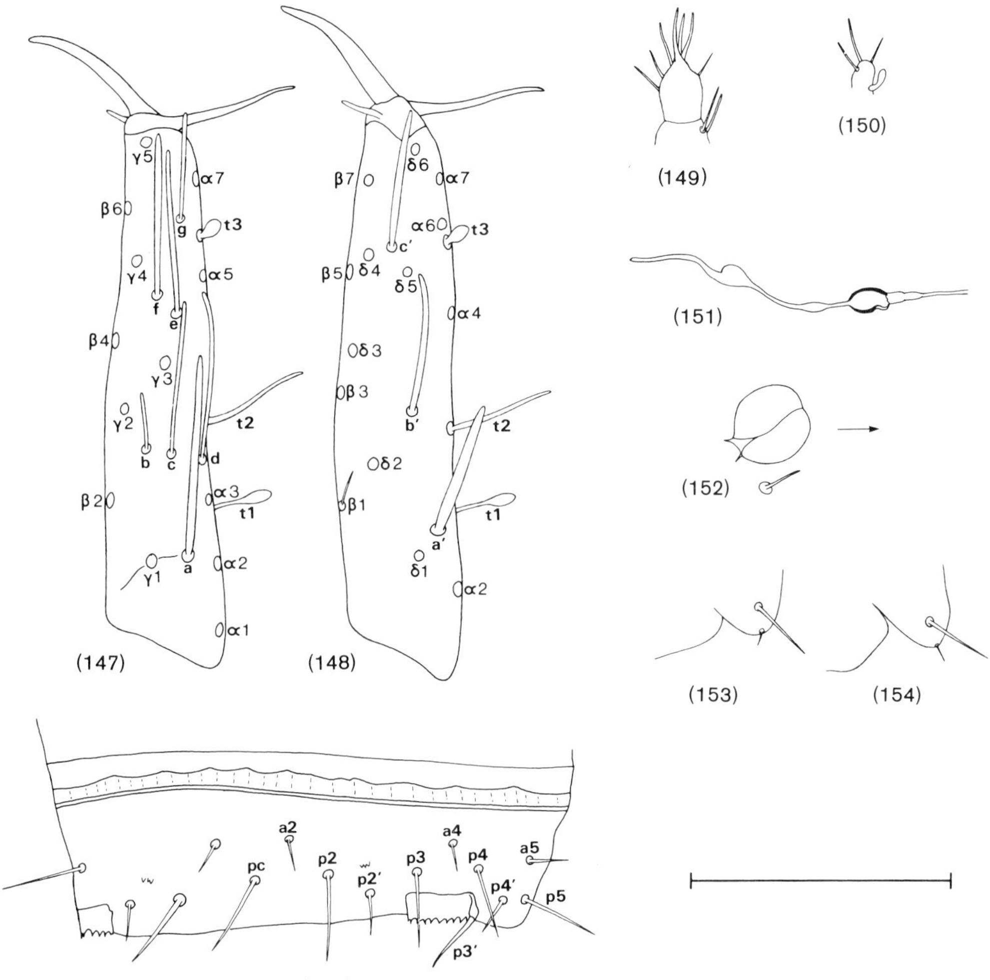

Figures 147–155. *Kenyentulus kenyanus*: 147, 148, foretarsus, exterior and interior view; 149, maxillary palp; 150, labial palp; 151, canal of maxillary gland; 152, pseudoculus; 153, 154, abdominal legs II and III; 155, tergal chaetotaxy of abdominal segment VIII. Scale line: 40 μm (153–155); 26 μm (147–152).

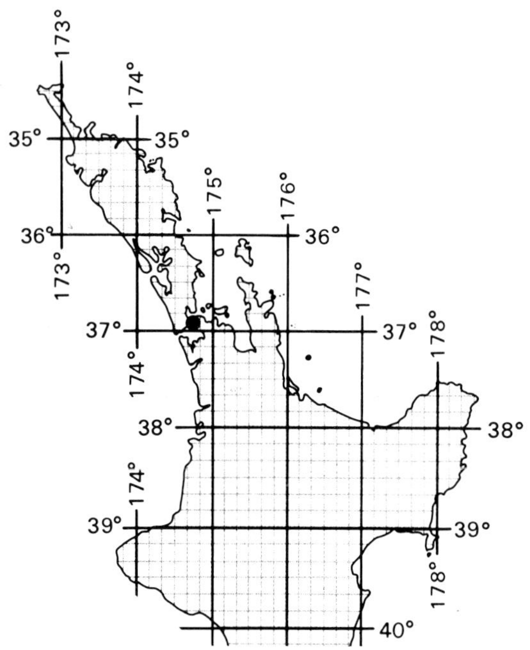

value to this character), and Bermuda (ZMKD; unpublished record). The distribution may be called pantropical, north and south of the equator.

The name clearly refers to the country of origin.

Genus *Berberentulus* Tuxen

Tuxen, 1963: 89. Type species *Acerentulus berberus* Condé, 1948, by original designation.

Acerentomidae with striate band reduced, the weak striae visible only in phase contrast. Labial palp reduced, with 2 or 3 setae. Second and 3rd abdominal legs with 2 setae, a long subapical one and a delicate median apical one. Sensilla *t1* on foretarsus claviform, *b'* present or absent. Sternum VIII without posterior setae.

Berberentulus nelsoni Tuxen

Figure 156–165

Tuxen, 1976: 427–428.
Berberentulus sp. (aff. *nelsoni* Tuxen) Ramsay & Tuxen, 1978: 601–603.

Length of body (extended) 1200 µm; length of foretarsus without claw 81 µm.

HEAD. Mouthparts of the common acerentomid type. Maxillary palp with a tuft and 2 very long, pointed sensillae (Figure 158). Labial palp reduced, with 3 setae and a sausage-shaped sensilla (Figure 159). Canal of maxillary gland with a heart-shaped calyx and a proximal part less than half as long as proximal branch of fulcrum (Figure 160). Pseudoculus circular (Figure 161); *PR* = 10.

THORAX. Foretarsus (Figures 156 and 157) with all sensillae present except *b'* on interior side; *t1* claviform; *t3* clumsy; *b* very long (reaching *f*), broad; *c*, *d*, *e*, and *f* slender, very long; *g* fairly short, exceptionally broad; *f* closer to *e* than to *g*; *a'* fairly long, sword-shaped; *c'* long, reaching base of claw; *b'* missing; *BS* = 0.5; *EU* = 0.1; *TR* = 3.6.

Chaetotaxy of metanotum: *a2*, *a4*, *M*; *p1*, *p1'*, *p2*, *p2'*, *p3*, *p4*, *p5*.

ABDOMEN. First abdominal leg with a terminal vesicle and 4 setae; 2nd and 3rd legs each with a long subapical seta and a short medial, apical seta (Figure 162).

Chaetotaxy of segments as in Table 14.

Terga II–V with *a1*, *a2*, and *a5* present; *a4* added to these on tergum VI; *a2*, *a4*, and *a5* on tergum VII; *p1'* present on terga I–VII; *p4'* on terga II–VII.

Striate band reduced, with no trace of striae. Comb VIII with 5 long, widely spaced teeth (Figures 163 and 164).

Squama genitalis of female (Figure 165) with short basal arms and acutely pointed acrostyli.

Type data. Holotype female, Brazil, São Paulo, Campinas, Fazenda Santa Genebra, March 1971, A. Pires Prado (Museo Zoologico, São Paulo).

Table 14 Abdominal chaetotaxy of *Berberentulus nelsoni*

	I	II,III	IV,V	VI	VII	VIII	IX	X	XI	Telson
Tergum	6/12	6/16	6/16	8/16	6/16	6/16	14	12	6	9
Sternum	3/4	3/5	3/8	3/8	3/8	4	4	4	6	6

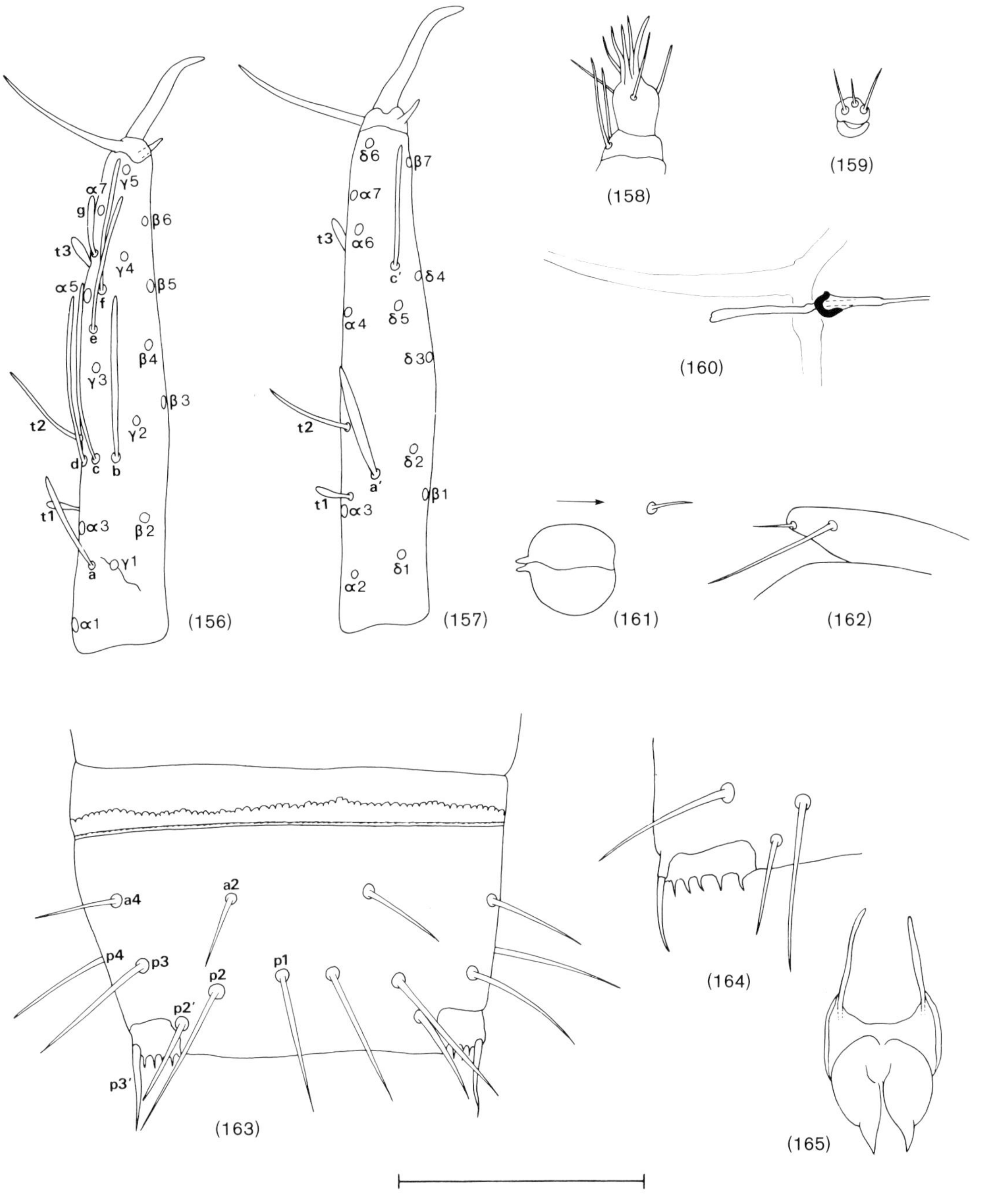

Figures 156–165. *Berberentulus nelsoni*: 156, 157, foretarsus, exterior and interior view; 158, maxillary palp; 159, labial palp; 160, canal of maxillary gland; 161, pseudoculus; 162, abdominal leg II; 163, tergal chaetotaxy of abdominal segment VIII; 164, comb VIII; 165, squama genitalis of female. Scale line: 40 μm (156, 157, 162–165); 26 μm (158–161).

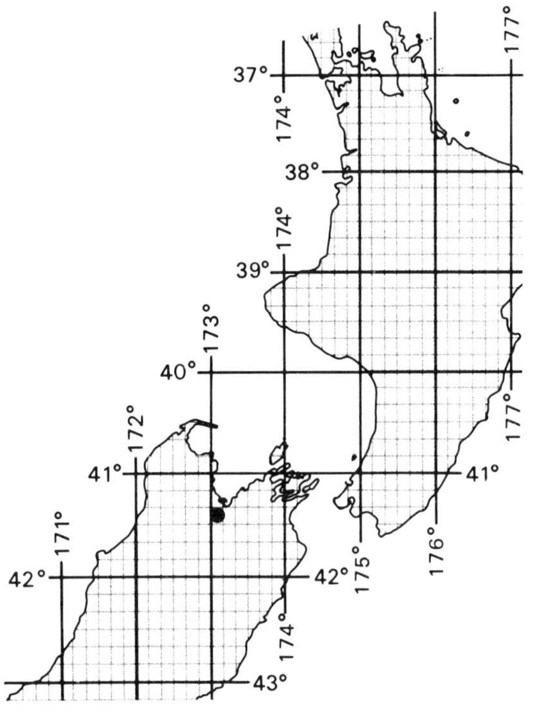

Material examined. Six non-type females (NZAC, ZMKD).

Kermadec Is / — / NN.

Collected in January, May–June, November, and December.

Taken from litter beneath *Rhopalostylus cheesemanii*, other mixed leaf litter, and in moss caps on pumice rocks.

Remarks. The first specimen of *B. nelsoni* examined from Kermadec Islands material (Ramsay & Tuxen 1978) was without forelegs, and thus unidentifiable; it was only the curious shape of comb VIII which led to a correct, if tentative, identification. It is remarkable that the species has been found together with *Acerentulus kermadecensis* (and no other) both on Raoul Island and in Pigeon Valley, Nelson.

This species has also been recorded from southern Brazil (São Paulo and Rio Grande do Sul).

The name was given in honour of Dr Nelson Papavero, of São Paulo.

REFERENCES

Bacetti, B.; Dallai, R.; Fratello, B. 1973: The spermatozoon of Arthropoda, XXII. The "12+0", "14+0" or aflagellate sperm of Protura. *Journal of cell science 13*: 321–335.

Berlese, A. 1908: Nuovi Acerentomidi. *Redia 5*: 16–19.

——— 1909: Monografia dei Myrientomata. *Redia 6*: 1–182.

Bernard, E. C. 1976: Observations on the eggs of *Eosentomon australicum* (Protura: Eosentomidae). *Transactions of the American Microscopical Society 95*: 129–130.

Bonet, F. 1950: Descripción preliminar de especies nuevas del género *Eosentomon* (Protura). II. El *E. pallidum* Ewing y sus especies afines. *Anales de la Escuela Nacional de Ciencias Biologicas 6*: 109–130.

Condé, B. 1948: Protoures de l'Afrique orientale britannique. *Proceedings of the Zoological Society of London 118*: 748–751.

——— 1952: Un Protoure de Nouvelle-Zélande. *Records of the Canterbury Museum 6*: 163–165.

——— 1955: Protura. *South African animal life 2*: 57–59.

Hennig, W. 1931: Insect phylogeny (translated and edited by A. C. Pont). Chichester, Wiley. 514 p.

Imadaté, G. 1965: Proturans-fauna of Southeast Asia. *Nature and life in Southeast Asia 4*: 195–302.

——— 1974: Protura (Insecta). *In:* Fauna Japonica. Tokyo, Keigaku Publishing Co. 351 p.

McColl, H. P. 1975: The invertebrate fauna of the litter surface of a *Nothofagus truncata* forest floor, and the effect of microclimate on activity. *New Zealand journal of zoology 2*: 15–34.

Nosek, J. 1973: The European Protura. Their taxonomy, ecology and distribution. With keys for determination. Geneva, Museum d'Histoire Naturelle. 345 p.

——— 1978: Protura: a new insect order record for Puerto Rico with discription of new species. *Journal of the Agricultural University of Puerto Rico 62*: 133–138.

Owen, H. G. 1976: Continental displacement and expansion of the Earth during the Mesozoic and Cenozoic. *Philosophical transactions of the Royal Society, London, A 281*: 223–291.

——— 1981: Constant dimensions or an expanding Earth? Pp. 179–192 *in:* Cocks, L. R. M. *ed.*, The evolving Earth. British Museum (Natural History) and Cambridge University Press.

Ramsay, G. W.; Tuxen, S. L. 1978: Protura (Insecta) from the Kermadec Islands, and a redescription of *Acerentulus nemoralis* from Argentina. *New Zealand journal of zoology 5*: 601–606.

Satchell, G. H. 1952: The order of Protura in New Zealand. *New Zealand entomologist 1*: 11.

Silvestri, F. 1907: Descrizione di un novo genere di insetti apterigoti, rappresentante di un novo ordine. *Bolletino del Laboratorio di Zoologia Generale e Agraria, Portici, 1*: 196–311.

Stevens, G. R. 1980: New Zealand adrift. Wellington, A. H. & A. W. Reed. xxii + 442 p.

Sturm, H. 1959: Die Nahrung der Proturen. Beobachtungen an *Acerentomon doderoi* Silv. und *Eosentomon transitorium* Berl. *Die Naturwissenschaften 46*: 90–91.

Tillyard, R. J. 1925: Primitive wingless insects. Part II: The orders Protura and Collembola. *New Zealand journal of science and technology 7*: 298–303.

Tuxen, S. L. 1931: Monographie der Proturen, I. Morphologie nebst Bemerkungen über Systematik und Okologie. *Zeitschrift für Morphologie und Okologie der Tiere 22*: 671–720.

——— 1949: Uber den Lebenszyklus und die postembryonale Entwicklung zweier dänischer Proturengattungen. *Det kongelige danske Videnskabernes Selskab, biologiske Skrifter 6, 3.* 49 p.

——— 1964: The Protura. A revision of the species of the world. With keys for determination. Paris, Hermann. 360 p.

——— 1967: Australian Protura, their phylogeny and zoogeography. *Zeitschrift für zoologische Systematik und Evolutionsforschung 5*: 1–53.

——— 1970 *ed.*: Taxonomist's glossary of genitalia in insects (2nd edition). Copenhagen, Munksgaard. 359 p.

——— 1976: The Protura (Insecta) of Brazil, especially Amazonas. *Amazoniana 5*: 417–463.

——— 1977: Protura (Insecta) of the New Hebrides. *Records of the South Australian Museum 17*: 299–307.

——— 1978a: Protura (Insecta) and Brazil during 400 million years of continental drift. *Studies on neotropical fauna and environment 13*: 23–50.

——— 1978b: The Protura (Insecta) of the Seychelles. *Entomologica scandinavica 9*: 251–263.

——— 1980: How to investigate phylogeny and how to use it, exemplified by Protura. Pp. 133–146 *in:* First international seminary on Apterygota, Siena, 13–16 September 1978. Siena.

——— 1981: The systematic importance of "the striate band" and the abdominal legs in Acerentomidae (Insecta: Protura). With a tentative key to acerentomid genera. *Entomologica scandinavica, supplement 15*: 125–140.

——— 1982: The Protura (Insecta) of Madeira. *Bocagiana, Funchal, 65*: 1–20.

——— 1984: *Brasilidia* and a new genus and species of Protura (Insecta) from Argentina. Together with an improved generic key to Acerentomidae. *Revue d'ecologie et de biologie du sol 21*: 283–295.

Tuxen, S. L.; Imadaté, G. 1975: The Protura of the Bismarck Archipelago and Solomon Islands. *Bulletin of the British Museum (Natural History), entomology 31*: 333–375.

Tuxen S. L.; Yin, W. 1982: A revised subfamily classification of the genera of Protentomidae (Insecta: Protura), with description of a new genus and a new species. *Steenstrupia 8*: 229–249.

Yin, W. 1981 (1983): On bionomics of *Sinentomon* (Protura) and its systematic position. *Contributions of the Shanghai Institute of Entomology 1981*: 161–170. [In Chinese, with English summary.]

—ꟹ—

TAXONOMIC INDEX

This index covers all the nominal taxa of Protura (and other groups of Arthropoda) mentioned in the text, regardless of their current status in taxonomy. Page numbers in bold type indicate the start of a major descriptive section. Those in italics indicate illustrations of the species concerned. The suffix "k" alludes to the page on which taxa are keyed out.

Australian Journal of Zoology

The *Australian Journal of Zoology* is a journal of international standard for the publication of the results of original research in all branches of zoology. Papers such as taxonomic revisions which cannot be accomodated by the journal because of their length are published as separates in the Journal's Supplementary Series. Critical review articles are also considered. All papers are refereed.

The Journal is one of nine published by the Commonwealth Scientific and Industrial Research Organization with the co-operation of the Australian Academy of Science.

All inquiries and manuscripts should be forwarded to the Editor-in-Chief, Australian Journal of Zoology, 314 Albert Street, East Melbourne, Victoria, 3002.

The Notice to Authors is published in the first issue of each volume. Copies are available on request.

Subscription Information

The *Australian Journal of Zoology* is published six times a year. Annual subscription rates for 1984 are as follows:

By surface mail	By airmail	Microfiche
$A120 with supplements	$A132 with supplements	$A40 with supplements
$A100 without supplements	$A112 without supplements	$A30 without supplements

---

Order Form

I wish to subscribe to the *Australian Journal of Zoology* and enclose cheque/money order.

Name ....

Address .... Post Code ....

Signature .... Date ....

Your order, accompanied by prepayment, should be sent to the 'Collector of Moneys, CSIRO', Editorial and Publications Service, 314 Albert Street, East Melbourne, Victoria, 3002 Australia.

TAXON :

15
14
13
12
11
10
9
8
7
6
5
4
3
2
1
MSL
–1

ALTITUDINAL DISTRIBUTION

(MSL, mean sea level) The range of altitude for each scale division may be set according to need, but should be in metric units.

OFFSHORE ISLANDS

- Kermadecs ☐
- Three Kings ☐
- Chathams ☐
- Snares ☐
- Bounty ☐
- Antipodes ☐
- Aucklands ☐
- Campbell ☐

This proforma data sheet may be photocopied as required; copyright is waived. Use the back for data such as host records.

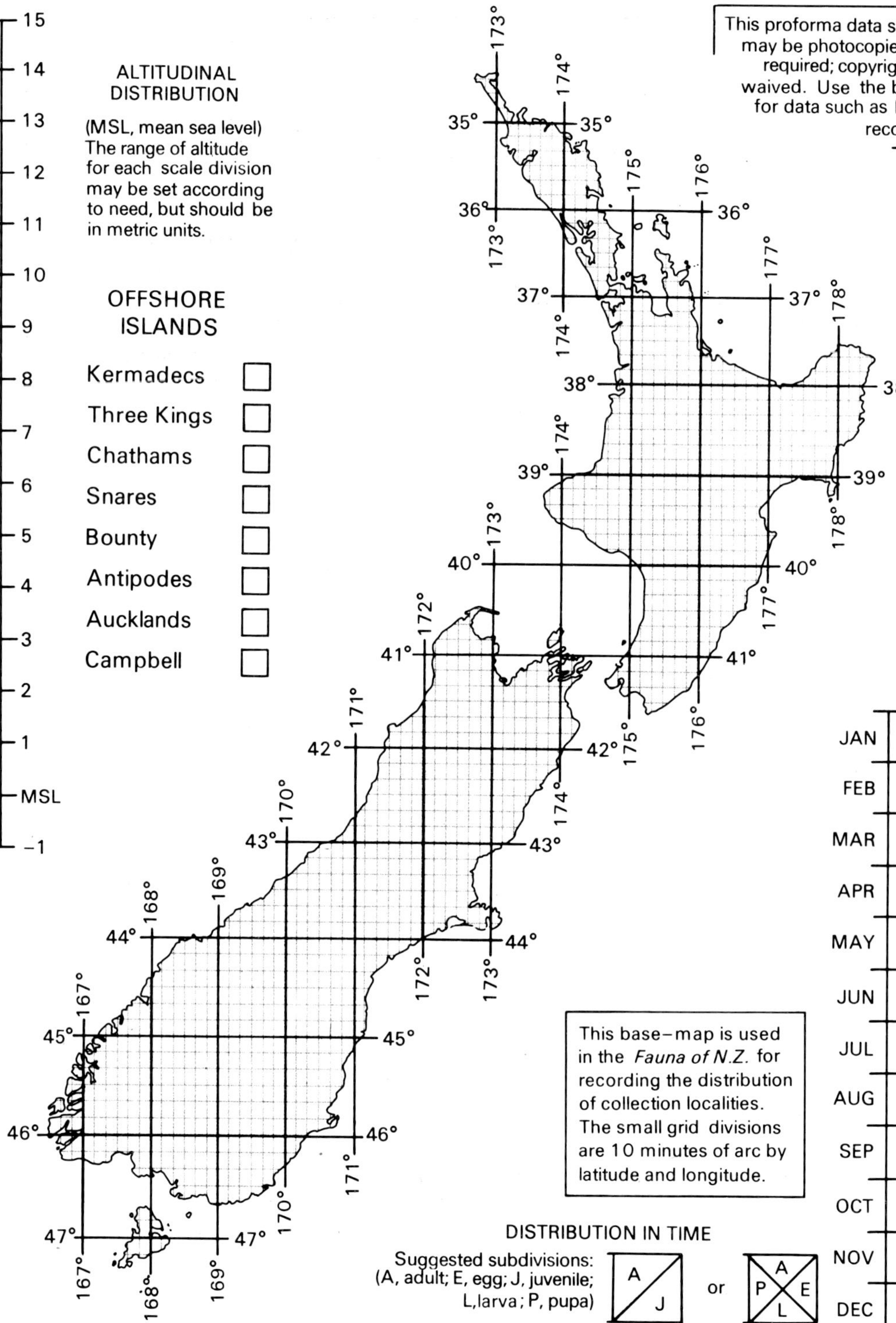

This base-map is used in the *Fauna of N.Z.* for recording the distribution of collection localities. The small grid divisions are 10 minutes of arc by latitude and longitude.

JAN	
FEB	
MAR	
APR	
MAY	
JUN	
JUL	
AUG	
SEP	
OCT	
NOV	
DEC	

DISTRIBUTION IN TIME

Suggested subdivisions: (A, adult; E, egg; J, juvenile; L, larva; P, pupa)

A / J or A, P, E, L

V. R. WARD, GOVERNMENT PRINTER, WELLINGTON, NEW ZEALAND—1985